When the
Universe
Took a U-Turn

When the
Universe
Took a U-Turn

B G Sidharth
B M Birla Science Centre, Hyderabad, India

World Scientific

NEW JERSEY · LONDON · SINGAPORE · BEIJING · SHANGHAI · HONG KONG · TAIPEI · CHENNAI

Published by

World Scientific Publishing Co. Pte. Ltd.

5 Toh Tuck Link, Singapore 596224

USA office: 27 Warren Street, Suite 401-402, Hackensack, NJ 07601

UK office: 57 Shelton Street, Covent Garden, London WC2H 9HE

British Library Cataloguing-in-Publication Data
A catalogue record for this book is available from the British Library.

WHEN THE UNIVERSE TOOK A U-TURN

ISBN-13 978-981-4277-81-5
ISBN-10 981-4277-81-9

Printed in Singapore.

Dedicated to the memory
of my father and mother

Preface

In the late sixteenth century, the German scholar Johannes Kepler inherited a huge amount of data about the positions of the planets, from his mentor Tycho Brahe who had just passed away. The motivated Kepler set upon improving the Greek model of the universe, armed with the latest figures. His technique was all too familiar as it had been tested for nearly two thousand years. He would have to add one more sphere or epicycle as it was called, to the orbit of a planet. Or he would have to slightly shift the centre of a circular orbit. Things like that. But try as he might, and that meant several years of frantic calculations, he could not bridge a tiny discrepancy between Brahe's observations and the model of the universe. The difference arose for the planet Mars, and it was miniscule, one might even say, not worth all these trials and tribulations. Kepler himself confidently declared that this minute discrepancy was really pointing the way to a reformation of astronomy. And so it did! The two thousand years old Greek model of the universe collapsed, when Kepler discovered that by using a slightly non circular orbit, an elliptical orbit in fact, the discrepancy vanished. This heralded the beginning of modern science.

Early in the last century, a similar situation prevailed. Breath taking progress had been made in science in the preceding centuries that had witnessed unparalleled contributions from the likes of Isaac Newton, James Clerk Maxwell and others. And yet there was a small something that didn't add up, a fly in the ointment. This was in the form of the results of some recent experiments which could be explained by the beautiful theories of physics and the universe, almost – but not quite. Once again science took a tumble. The new and minute discrepancies gave birth to the special theory of relativity, the general theory of relativity and Quantum theory, which undoubtedly were the intellectual peaks of the last century. They have gone a long way in shaping and reshaping our concepts of the universe and indeed technology which touches our lives. At last after millennia of scientific and intellectual quest, everything seemed to be falling in to place.

However, as the last century drew to a close certain discrepancies surfaced. Some of them were miniscule by any standards. For instance the mysterious particle called the neutrino was supposed to be massless. It turned out to have the slightest of masses, much, much lighter than the lightest known particle, the electron.

The realization has been creeping into the scientific community, that we are missing out on something. The awe inspiring, mind blowing contributions of the last century may not be the last word. These concepts were bizarre and upturned the cosy picture of the universe which was inherited from the time of Newton. But perhaps the radical ideas left behind by twentieth century physics are no more than a fore taste. A new cosmos seems to be emerging which is even more bizarre, even more crazy than we could ever imagine. We may well be at another dramatic turning point in our intellectual evolution and conception of the universe, confronting unexpected, even unimaginable concepts of spacetime and the universe. The story is still unfolding.

I have tried to capture the mood and spirit of this new paradigm shift in the following pages. Undoubtedly, in the course of simplification, to reach out to a larger audience, precision would be lost. This is because it is not possible to go into the merciless technical detail. In the process, to help the cause, I have here and there repeated certain ideas and concepts, and have almost always tried to keep the references at a popular level. I am grateful to Prof. Walter Greiner of the Frankfurt Institute of Advanced Studies for useful advice and also my secretary Mrs. Y. Padma and my wife Dr. B.S. Lakshmi for invaluable help in the preparation of the manuscript. My thanks are also due to Prof. K.K. Phua and Lakshmi Narayan of World Scientific, for unstinted cooperation.

B.G. Sidharth

Contents

Preface   vii

List of Figures   xi

1. The Lord is subtle, but defies commonsense!   1

2. You can be younger than your grandson   11

3. God does not play dice, or does He?   31
   - 3.1   The Classical Catastrophe 31
   - 3.2   Particles Disguised as Waves 37
   - 3.3   The Paradox of Reality 47

4. Time is running backwards isn't it?   57
   - 4.1   Dealing With Billions 57
   - 4.2   Time, the Eternal Enigma 66

5. On a collision course   75
   - 5.1   Bits of Atoms 75
   - 5.2   Fundamental Forces 86
   - 5.3   Quantum Gravity 93
   - 5.4   Strings and Loops 100
   - 5.5   A Critique 109
   - 5.6   Smashing the Atoms 122

6. Law without law   127
   - 6.1   The Perfect Universe 127

6.2   The Lawless Universe   138

7.   When the universe took a U turn   151

7.1   The Exploding Universe   151
7.2   Birth Pangs of the Universe   152
7.3   The Infant Universe   154
7.4   A Dark Mystery   156
7.5   The U turn   158
7.6   A Darker Mystery   160
7.7   Other Footprints   164
7.8   A New Dawn   168
7.9   Blow out Infinity   171

Bibliography   177

Index   181

List of Figures

1.1   Greek epicycles and effective motion   5
1.2   An elliptic orbit   7
1.3   The stone tends to rush off along the tangent if the string snaps   8

2.1   Foucalt's pendulum   12
2.2   Stationary and moving charges   15
2.3   Boy running with and against the ship's motion   16
2.4   Light Clock   21
2.5   Ball falling in a stationary elevator   24
2.6   Ball floating in an accelerated elevator   24
2.7   The deviation and bending of light   26
2.8   The bending of light observed during a total solar eclipse . . .   28
2.9   Different speeds of a falling object at different heights   30

3.1   Rutherford's Gold Foil Experiment   34
3.2   Standing Wave   37
3.3   The addition of waves   38
3.4   Interference of waves   39
3.5   The uncertainty spread of a particle   41
3.6   The Stern-Gerlach Experiment   42
3.7   The double slit experiment   46
3.8   Schrodinger's Cat   49
3.9   EPR paradox experiment   51

4.1   Diffusion of molecules   61
4.2   The universe as an apple   67
4.3   Projectiles and Satellite   69
4.4   A random walk   72

Chapter 1

The Lord is subtle, but defies commonsense!

"The way that can be walked is not the perfect way, the word that can be said is not the perfect word"

-Lao Tse, c. 4th Century B.C.

Malcom, an American astrophysicist was peering into his computer screen carefully examining the tons of data beamed back by NASA's Gamma Ray Large Array Space Telescope, GLAST. This had been blasted off in mid 2008 to relay information on Gamma Rays being belched out from the deepest recesses of the universe. Was there a clue that could end the reign of Einstein's Special Theory of Relativity, a theory that had withstood a century of scrutiny?

A continent away, Jacob, a particle physicist in Europe was meticulously computing, using the equally voluminous data being churned out by the recently kick-started Large Hadron Collider (LHC), in CERN, Geneva. Was there a faint footprint of the Higgs particle that had been eluding detection for over four decades? Was the standard model of particle physics erected more than three decades earlier, the last word?

The year, 2010. Both the GLAST and the LHC are multi-billion dollar contraptions, the culmination of years of planning and labor, eagerly awaited for several years by physicists the world over. These would test two of the greatest intellectual theories fashioned out by the human mind in the last century of the millennium, the Theory of Relativity and Quantum Mechanics.

The story begins some 12000 years ago. The Earth was a frozen planet, much of it a freezing white sheet of ice. Human beings lived in caves and caverns, prowling and wandering like polar bears. They wore furs and used crude stone tools in their efforts to hunt down wild animals for food. But

1

already the last of the great ice ages had begun to thaw and a new lifestyle was beginning to emerge. Agriculture. People now had to settle down in habitats to tend to their crops and domesticated animal herds.

As people looked around at the universe - the earth, sky and nature, in bewilderment, they discovered that oftentimes hostile nature could also be an ally in the struggle for survival. For example, they could use clubs or fire and later the wheel.

The advent of agriculture was a major departure in life style from the hunter gathering ways and days. People soon discovered that the drama in the heavens - the rising and setting of the sun, the phases of the moon and the progression of the seasons provided a clock cum calendar so essential for the new science of cultivation. Roughly, the starry sky was the dial of a clock, while the sun and moon were like the hands that indicate the passage of time.

The heavens threw up the different units of time. For instance the sun rises, sets and rises again. This gives birth to the 'day'. The moon passes through its various phases - from full moon through crescent moon back to full moon. This gives rise to the 'month'. Then, the sun appears to have another motion apart from rising and setting. This is a slow west to east drift through the stars. This means that exactly at sunset each day, we see different groups of stars or constellations as we call them, in the west. But after a year, we return to the same pattern of stars at sunset. These basic units of time were used for framing the earliest calendar, which guided ancient agricultural activity.

However as the precision of observations increased, crude calendars gave way to more precise ones in order to keep a track of the seasons, so essential for sowing, harvesting and other agricultural activities. Briefly, the problem which ancient man faced was the following. The month consisted of twenty nine and a half days, while the year consisted of three hundred and sixty five and one fourth days. So twelve months would be three hundred and fifty four days, some eleven days less than the true year. You might say, so what? But this difference of about eleven days would pile up. After about sixteen years, the difference would be about six months – and winter would be summer and summer in winter. A mess!

The ancient Indian calendar, perhaps the precursor of the Sumerian and later calendars overcame this mismatch by using a simple technique. As every three years of twelve months would lead to a shortfall of some thirty three days, an extra month was added every three years, a leap month.

Of course this scheme had to be tweaked even more to bring into step the month and the year.

The point is that the months are computed using the moon. This is very accurate because we can see the backdrop of stars when the moon is present in the sky. However the year is dictated by the sun – for instance the interval between the time when the length of the day (sunrise to sunset) exactly equals the length of the night to the next such occasion. All this happens at the vernal equinox. Similarly there is the autumnal equinox or the winter solstice signalling the longest night and the shortest day and the summer solstice where the opposite happens. It is the year which is important for agriculture, because it is so intimately connected with the seasons. What we are trying to do is to match the months of moon with the year of the sun. Small differences will have to be eliminated. For instance even if we add one extra month every three years, there is still a residual difference of a few days. Such a calendar is called a luni solar calendar and has been the hallmark of ancient agricultural civilizations.

However, one could just use the month as a unit and try to fit in the three hundred and sixty five and one fourth days of the year. Twelve months. This is essentially what has been done over the past two thousand years. Our modern calendar for instance manages to fit in three hundred and sixty five days within the twelve months. The one fourth day could be adjusted by having three years with three hundred and sixty five days and a fourth year, the leap year with an extra day. Again this broad scheme had to be fine tuned to make it precise.

As thought evolved, people began to try to understand and explain what they saw. In the process they built models, that is described new phenomena in terms of concepts or vocabulary they already knew. This is what every growing child does. The pattern has been going on ever since.

The earliest known model builders were the composers of the Rig Veda, several thousand years ago [1]. With amazing insights, they could guess that the apparently flat earth was actually round, describing the earth and sky as two bowls. They went on to describe the sun as a star of the daytime sky and even asked, how is it that though the sun is not bound it does not fall down? And so on and so on. This is an example of a pattern that has repeated itself over the ages: the breakthroughs have been counterintuitive, that is, have defied the apparent dictates of commonsense.

As we will see over and over again, the universe is not what it appears to be. The answers may go against our intuition which is based on our experience. An example of an intuitive model of the universe would be that of the ancient Egyptians for whom the sky was a ceiling, supported at the extreme ends by mountains. It is difficult to break out of the intuitive mold. Such a departure is often forced upon us, due to the utter inadequacy of old concepts to describe new phenomena.

Perhaps the earliest model of what we today call microphysics or atomic or subatomic physics was proposed by the ancient Indian thinker Kanada who lived around the seventh century B.C. For him the universe was made up of ultimate sub constituents which were in perpetual vibration. Later, Greeks also had an atomic theory, which they may or may not have acquired from India. But there was a crucial difference. Their atoms were static. These concepts were brilliant, but they were much too counterintuitive, much too fanciful to be accepted wholeheartedly. How could smooth marble, for example, be composed of separate discrete bits? They remained on the fringes of speculation and philosophy for over two thousand years.

The insights of the Vedas were camouflaged in allegory and this oral tradition was lost over the millennia. Our legacy of modern science came from the more intuition and experience based Greeks who were expert geometers. They built up over a few centuries, an even more complex cosmic scheme in which the flat earth was at the center, surrounded by a series of transparent material spheres to which the various heavenly objects like the sun, moon, planets and stars were attached.

Why the material or crystalline spheres? Well, they were necessary, for, otherwise they would have had to explain why the moon doesn't crash down on to the earth, for example. Newton was not the first man to notice that the apple falls down!

These were spheres because Plato the Greek thinker, had preached that the circle (or sphere) was a perfect object, due to its total symmetry. Plato and his followers were obsessed with this notion of symmetry and perfection, with origins in geometry. His school, for example proclaimed that no one was allowed in, who did not know geometry. This legacy has come down over the ages – even Einstein swore by geometry. The moon is round, and so is the sun. In a perfect world therefore, everything had to be circular or spherical. The modern word "orbit" comes from the Greek word, orb, for circle. Furthermore these spheres would have to be rotating – otherwise how could we explain the heavenly drama such as the rising and setting of the stars.

As the observations sharpened, the above simple model, first put forward by Anaximenes around 500 B.C. needed modifications [2]. For instance the center of a sphere would not coincide exactly with the earth, but rather would be eccentric, that is, slightly away from it. Then the spheres themselves had to carry additional spheres called epicycles, themselves spinning and the objects were placed on top of the epicycles. Later, around 100 (A.D.), Ptolemy the Librarian of the Great Library of Alexandria compiled all this knowledge in two astronomical treatises only one of which, the great Astronomer or Al Magest survived. The Ptolemaic universe was a complicated tangle of such spheres and epicycles and epi-epicycles, undergoing complex circular motions. It was a geometrical wonder though. (Cf.Fig.1.1).

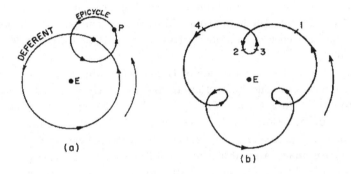

Fig. 1.1   Greek epicycles and effective motion

These basic ideas survived for nearly two thousand years, till the time of Copernicus. Nicolaus Copernicus was born in Torun, today in Poland, to a family dealing with copper. By the time he was forty, he settled down in Frauenberg, Germany. In the sixteenth century, in a commonsense defying leap of faith, he put the sun rather than the earth at the center, though much of the Greek baggage like transparent material spheres and the tangle of epicycles remained. Nevertheless it was a dramatic, even absurd idea. Imagine the solid earth with the trees and buildings hurtling through space! Another contra idea that was smuggled in was that the everywhere flat earth was actually round.

Though such contra ideas had been put forward by a Greek, Aristarchus, more than two thousand years before Copernicus, they were rejected even by the scholars. They argued decisively over the centuries that earth could not be traveling or rotating.

By now there had been a major development. Unfortunately the Greeks did not know of the positional system of notation and so their numbers were the clumsy so called Greek numerals. These were not amenable to algebraic computations. However the positional system of notation including the all important concept of zero were already well established in India over two thousand years earlier. No wonder that the ancient Indians were way ahead in computational techniques. For example Brahmagupta in the seventh century was already using second order differences in his interpolation formulas. So also Aryabhatta was using Trignometrical tables and computations around the same time. Moreover, calculus, an indispensable tool, was being used for planetary calculations at least six centuries before Newton [3].

In the eighth century the founder of Baghdad Khalif Al Mansoor invited Indian scholars, and one in particular, Kanka of Ujjain in Central India acquainted the Arabs with not only the decimal system but also Brahmagupta's computational techniques. The decimal system itself reached Europe through Spain as late as the twelfth/thirteenth century. Two mathematicians, Fibonacci and later Abelard of Bath learnt of this new technique and it soon caught on. Copernicus [4] and later astronomers were armed with this new and powerful tool of mathematics.

But the bold work of Copernicus did not go far enough. Early in the seventeenth century, Kepler a young German scholar, noticed that the Copernican model differed from observation by just a whisker, eight minutes of arc, for the orbit of Mars. Kepler had inherited the meticulous observations of his mentor Tycho Brahe [5], and a lesser mortal would have put down this minor discrepancy to an error in observation. On the contrary, Kepler was convinced that the observations were correct and that the discrepancy pointed to a reformation of Astronomy, as he put it.

Kepler in another counterintuitive leap of faith proposed his first two laws of planetary motion in 1609. Some years later the third law followed. Crucially the orbits were ellipses. This went blatantly against Plato's picture of perfection that had survived two thousand years. With a single ellipse Kepler could wipe out the minute discrepancy between theory and observation, for the planet Mars. There were different ellipses too for other planets. The tangle of crystalline material spheres which had withstood two millennia of scrutiny, finally came crashing down. (Cf.Fig.1.2).

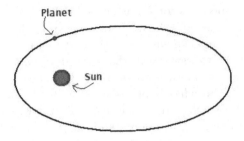

Fig. 1.2   An elliptic orbit

Kepler's third law related the sizes of the orbits of the planets to the time taken by the planets to go round the sun couched in a precise mathematical language. Indeed this was the beginning of modern science.

In hindsight, what the Greeks had tried to do was, approximate a simple ellipse by a series of complicated "perfect" circles. The larger implication of this minute correction was this: The ellipses destroyed the crystal spheres holding up the sun, moon, planets and the stars. And – the age old question once again returned to haunt. Why don't the moon, the planets and so forth crash down?

This question was answered by the brilliant British scientist, Sir Isaac Newton. He first crafted the laws of mechanics which were based on ideas developed a little earlier by the Italian scholar Galilei Galileo. Next he introduced his Theory of Gravitation. Kepler's purely observational laws could now be explained from theory.

Newton's answer to the age old puzzle was actually opposite to what we might naively expect. There was no counterbalancing force to halt the planets from plumetting down. Such a force, as we will see, came from

Einstein much later. Rather it was the attractive gravitational force which held aloft these objects in their orbits! How could that be?

First we should realize that for thousands of years people had taken it for granted that a force was needed only to move an object. Even Aristotle the famous Greek had said so. On the contrary Galileo and Newton noticed that without any force an object could be stationary or it could even move with uniform speed. A force was needed only to change the speed or direction of motion. This was the content of Newton's law of inertia.

So, let us consider the moon going round the earth for example. If suddenly the gravitational force was switched off, what would happen? Rather than plummet, the moon would rush off in a straight line at a tangent to its orbit. This is similar to what happens when a stone whirling at the end of a string is suddenly released (Cf.Fig.1.3). The attractive gravitational force would actually tug at the moon, pull it in and barely keeping it in its orbit.

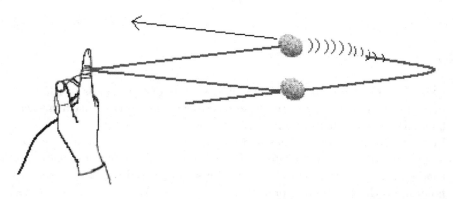

Fig. 1.3   The stone tends to rush off along the tangent if the string snaps

The beautiful edifice of Newtonian mechanics which demolished the Greek crystal palaces dominated the scientific scene for a few centuries - less than a tenth of the life of the Greek model itself. There was an absolute space, while time was separate. Space was a container a platform on which actors like matter, force and energy played their roles. Time too was absolute. The present moment "now", meant the same for a person on the street or an extragalactic being. A meter or a second were universal. All this is, of course "commonsense". Albert Einstein would say, the Lord is subtle, but he is not malicious. The Lord though defies "commonsense" as we will discover.

Chapter 2

You can be younger than your grandson

"I want to know how God created this world. I am not interested in this or that phenomenon, in the spectrum of this or that element. I want to know His thoughts, the rest are details."

A. Einstein

The laws that emerged from the work of Galileo, Kepler and Newton constituted what we today call mechanics. The question whether these laws are valid for all observers or for some privileged observers– and if so to whom – troubled Newton. This question is not immediately obvious. For example if we say that the laws are true as seen from the earth, then the question would prop up, what about the laws as seen by some one on the planet Jupiter or elsewhere? The joke is that Newton's laws are not true on the earth itself, as we will see!

In the Greek model the earth was at the centre and stationary and such a question did not really bother anyone. All that we had to say was that the laws were true for any person who was at absolute rest like the earth itself. This would not be the case if the earth were hurtling through space.

Newton's answer to the question was that the laws would be true for an observer for whom the distant stars were fixed, that is stationary. By this yardstick the laws would not be true on the earth, as noted. This is because the distant stars rise and set – they are not stationary. Today we would say that Newton's laws could be observed by one for whom the distant stars would appear stationary for example, a futuristic astronaut in a voyage towards some distant star. Such an observer, for whom the law of inertia and other laws hold is called an inertial observer.

The laws would not be true on the earth because a person on the earth is not inertial. This is brought out quite dramatically by the pendulum

experiment performed by the French scientist, Leon Foucalt in Paris in the
nineteenth century (Cf.Fig.2.1). He observed a long swinging pendulum.
Such a pendulum should oscillate only in a single vertical plane under the
influence of the earth's gravity. This plane however should not change, if we
are indeed inertial observers. That is because there is no tangential force
working on the pendulum bob, perpendicular to the vertical plane of its
oscillation. Let us say that there is no horizontal force acting on the pendulum. However Foucalt demonstrated that the plane does indeed change,
in fact it rotates once in a little under thirty three hours [6]. This figure
varies from place to place.

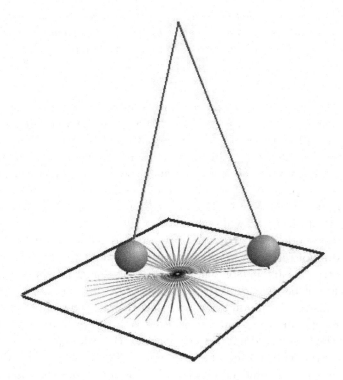

Fig. 2.1   Foucalt's pendulum

We can easily guess that this is because, the earth below the Foucalt
pendulum behaves like a rotating platform. At the North Pole itself, when

you are on the earth's axis of spin, the plane would rotate once in twenty four hours. That is for an observer not in contact with the earth, the pendulum would continue to swing in the same plane according to Newton's laws, but for an earth observer who is unaware of the fact that the earth below him is rotating, the plane of swing appears to rotate in the opposite direction. Apart from this however the earth approximates an inertial frame in which Newton's laws are valid.

Now let us suppose that we are in a spaceship – let's call it Space Lab – rushing through space at a uniform speed, as seen by someone on earth. If we perform experiments in our Lab, then just like Newton, we would soon enough rediscover all the laws of mechanics. It is irrelevant that we are moving (with a uniform speed). From this point of view we have as much right as any to call ourselves inertial observers. (As we will see soon, this would not be the case if our spaceship was accelerating). That all "inertial" observes have the same laws of mechanics is called the Principle of Galilean Relativity.

Apart from mechanics another curious phenomenon that was known for at least two thousand years was that of electricity–this was the strange behavior that was witnessed when some substances like amber were rubbed vigorously. This experience was formalized at the end of the eighteenth century: The electric charges exerted forces much like Newton's gravitational force. In fact the French experimenter, Coulomb discovered, using what is called a torsion balance that the electric forces mimic the gravitational force with the same inverse dependence of the square of the distance. The gravitational force depended directly on the masses of the two objects, while the electric force depended on their charges. From this point of view, we could even call the mass, the gravitational charge.

However there were two crucial differences. Firstly the gravitational force is always attractive, whereas the electric force came in two varieties – as an attraction between unlike charges, that is a positive charge and a negative charge, or as a repulsion between like charges. Secondly, while in principle the gravitational force could vary continuously in strength, with the mass of the objects, the electric force came with discrete charges. Thus there could be one, or two or three ... charges with nothing like one and a half charge. In contrast the mass, which plays the role of charge in gravitation, can be anything.

Shortly thereafter, early in the nineteenth century, Hans Christian Oersted performed an experiment which was sensational. By now it was known that an electric current was nothing more than a stream of moving charges, much like a stream of water. The big surprise was that such an electric current could deflect a magnetic compass. Though magnetism had been known for many centuries, its link with electricity came as a big big surprise. Other experiments soon followed.

Ampere for instance showed that if electric currents were passing through two parallel wires, there was a new force between these wires which resembled magnetic effects. There was a twist in the plot here. The theory of electric charges so far was on the lines of Newtonian mechanics and gravitation. Now it appeared that the force between wires was perpendicular to the direction of motion, something which we do not encounter in Newtonian mechanics.

Around the middle of the nineteenth century it was realized that electricity and magnetism were not the two different phenomena they were supposed to have been. Rather they were intimately interlinked. This was called electromagnetism. Only when the electric charges were moving slowly or were at rest could we speak about electricity as separate from magnetism. The British physicist James Clerk Maxwell worked out the detailed mathematical behavior of this new force of electromagnetism. It appeared that a moving charge would emit radiation that would travel through space and impinge upon other charges, rather like the ripples that glide across the surface of a pond when a stone is thrown on it. Further the speed with which such so called electromagnetic waves or ripples traveled, turned out to be miraculously the same as the speed of light. Today our entire infrastructure of telecommunications rests on this electromagnetic theory. The question was now being asked, was light itself some form of electromagnetic radiation?

What you might well wonder is, is there a principle of relativity for the new electromagnetic laws as for Newton's laws of mechanics? In other words could we say that the astronaut in the uniformly moving Space Lab, but otherwise cut off from the outside world discover the same laws as Maxwell did on the earth? At first sight the answer would appear to be easy– No. This is because if there were two charges at rest within our Space Lab, then we who are in it would observe that they interact according to Coulomb's law for electric charges. But, from the earth perspective, the two charges are moving with the Space Lab and so constitute currents. These interact in a completely different way, via Ampere's laws. (Cf.Fig.(2.2)).

So our description would be very different from that on the earth. But, as we will see soon, there was a mind blowing twist.

Newton's concept of time had been pretty much the same as that coming

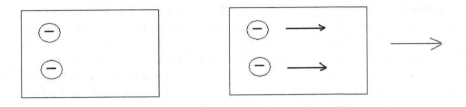

Fig. 2.2   Stationary and moving charges

down to us over the ages. In particular an instant of time had the same meaning for observers anywhere in the universe. So if a star exploded now, at the same time as a bird alighting on the branch of a tree, this statement was true universally. There was no way a person in another galaxy, for example would see this event as a star exploding first and several minutes later a bird alighting on the tree, or the other way round. In other words time was absolute. Not that there was any experiment that definitely showed the absoluteness of time. Nevertheless this had been taken for granted from time immemorial.

However towards the end of the nineteenth century some experiments threw up troublesome results. These were a series of experiments carried out by Albert Michelson and Edward Morley in the US. But the story really began in 1675 [7]. Till then, people had taken it for granted that light reached one point from another instantly, that is travelled with infinite speed. That year, a Danish astronomer, Olaf Romer, brilliantly conjectured that light travels with a finite speed. This conjecture was inspired by the invention of telescopes some sixty five years earlier. Around the mid seventeenth century a number of astronomers observed the eclipses of the satellites of the planet Jupiter. These eclipses were generally observed before and after Jupiter's opposition, when Jupiter and the Sun were on opposite sides of the earth as those were favourable times. However the eclipse times did not match those when Jupiter was near a conjunction, that is on the same side of the earth as the Sun. There was a difference of more than ten minutes. Romer speculated that this difference was due to the different distances of Jupiter on the one hand and the fact that light travelled with a finite velocity on the other.

To understand the Michelson-Morley experiment, imagine that we are on the seashore watching a ship sail by, let us say at a speed of forty kilometers per hour. If there were a boy on the deck running in the same direction with a speed of twenty kilometers per hour, to us the boy would appear to be moving with a speed of forty plus twenty, that is sixty kilometers per hour. In case the boy was running in a direction opposite to that of the ship, we would say that he was actually moving with a speed of forty minus twenty that is twenty kilometers per hour. (Cf.Fig.2.3).

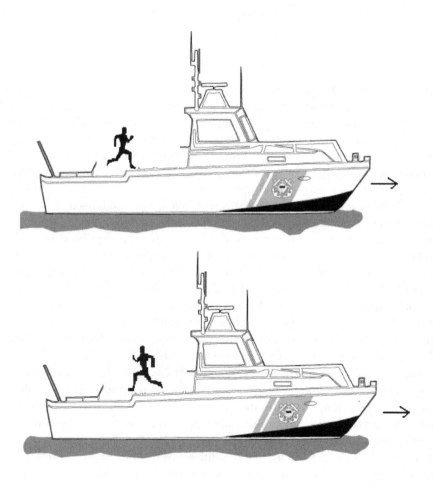

Fig. 2.3   Boy running with and against the ship's motion

This is commonsense and it is also the result given by Newtonian mechanics.

Now let us replace the moving ship with the earth. Remember in its own right the earth is a space ship, rushing around the sun with the enormous speed of about thirty kilometers per second which we will denote by v. Let the boy be replaced by a ray of light which moves with a speed c which is ten thousand times greater than v. By the above commonsense or Newtonian mechanics reasoning, we will conclude that, a ray of light traveling in the direction of the earth's motion towards a mirror would move with a speed c - v as the mirror rushes away from the ray due to the earth's motion. After reflection from the mirror it would return with a speed c + v because we on the earth are rushing to receive the ray. If on the other hand, light were traveling towards the mirror in a direction perpendicular to that of the earth's motion, its up and down journey would be carried out at the same speed c. The point is, to simplify matters, there would be a minute difference between the return times of light in these cases, even if the distance covered would be the same.

Michelson and Morley devised an experiment which was sensitive enough to detect this difference, even though it would be very minute [8]. However try as they would, such a difference did not show up. This was a huge embarrassment and puzzle for physicists. It upturns our logic. It is as if the boy on the ship would appear to be moving with the same speed – up or down. A couple of way out hypotheses were put forward to explain the absence of this miniscule difference, or in other words why there was no difference in the time of arrival of light in either case.

One of these was put forward by the great French mathematician and physicist Henri Poincaré. His ingenious idea was that we had taken it for granted that time intervals measured by clocks moving along and perpendicular to that of the earth's motion were the same: If a clock were traveling along the earth's direction of motion and another perpendicular to it, one second on either clock would have the same meaning. But suppose a second along the direction of the earth's motion was actually a wee bit longer than that in a perpendicular direction? Sounds preposterous doesn't it? In other words he assumed that time would dilate with motion. Poincaré could manipulate such a minute difference or dilation of time to deduce that there would be no difference in the arrival times of light, which would of course explain the results of the Michelson – Morley experiment.

A renowned physicist, H.A. Lorentz of Holland, as also the Irish physicist G.F. Fitzgerald came up with a different though equally brazen explanation. They noticed that we had taken it for granted all along that the length interval, for example a foot rule in the direction of motion was the same as a foot rule perpendicular to the direction of motion. This of course was commonsense and time tested for thousands of years. But suppose, however improbable it may seem, a foot rule in the direction of motion were less than one in a perpendicular direction, though by a very minute amount? This would also explain the arrival time puzzle.

It was left to find out which of the two explanations would agree with other experiments. Interestingly in hindsight, if a hypothetical experiment were done to determine if Poincaré was right or Lorentz, such an experiment would have shown that both were correct, thus compounding the puzzle!

It was at this stage that Albert Einstein, then a lowly patent clerk in Switzerland wrote an article with a rather uninspiring title, "On the Electrodynamics of Moving Bodies". Einstein was a run of the mill student, who in fact was forced to leave his school in Munich [9]. Thanks to his friend, the mathematician Marcell Grossman, he could study at the Swiss Federal Institute and moreover got his patent clerk job. In this paper Einstein assumed, that is took as postulates two mutually contradictory statements. The first was that the laws of physics, including electrodynamics would be the same for all observers moving with a constant velocity with respect to each other.

He wrote, "Examples of this sort, together with the unsuccessful attempts to discover any motion of the earth relative to the "light medium," suggest that the phenomena of electrodynamics, as well as of mechanics, possess no properties corresponding to the idea of absolute rest. They suggest rather that ... the same laws of electrodynamics and optics will be valid for all frames of reference for which the equations of mechanics hold good. We will raise this conjecture (the purport of which will hereafter be called the "Principle of Relativity") to the status of a postulate, and also introduce another postulate, which is only apparently irreconcilable with the former, namely, that light is always propagated in empty space with a definite velocity c which is independent of the state of motion of the emitting body." Now the first postulate as we saw, was correct for Newtonian mechanics, but we argued that it would be false for electromagnetism. It would lead precisely to the conclusion that the Michelson-Morley experiment should show the difference between the arrival times of light referred to.

The second postulate in a sense contradicted the first "impossible" hypothesis. This was that light would travel with the same speed for all such observers, that is the $c + v$ and $c - v$ speeds would be the same. This of course would contradict the commonsense results or in more technical language the relativity of Newtonian mechanics according to which light would have different speeds going up and coming down. This was an equally impossible hypothesis.

But Einstein argued that if we put these two contradictory postulates together, then the results of all experiments could be explained! In that 1905 paper, as we saw Einstein declared, "... the same laws of electrodynamics and optics will be valid for all frames of reference for which the laws of mechanics hold good". That is of utmost importance. In any case, as the Late Nobel Laureate Professor Abdus Salam would say, "Experiment is at the heart of physics". You can put forward bizarre hypotheses, but at the end of the day, the results must be in tune with observation. If all the crows we see are white, then so be it!

But how could this be? Einstein explained that the exotic hypotheses of both Poincaré and Lorentz were correct. The contradiction appears when we carry over the baggage of Newton's ideas of space and time, according to which space intervals and time intervals, as we saw earlier, had an absolute meaning, independent of who was measuring them. Einstein argued, experiment showed that this was not the case, and experiment is the key that lays down the rules, not some preconceived "commonsense" notions that we may have, which are really experience based prejudices masquerading as commonsense. In other words our concepts of space and time were based on preconceived notions, unrelated to the new experimental reality. Einstein brought in a physical concept of space and time, based on experiments. Of course we do not need to be too critical of ourselves because the differences which Einstein brought in are minute indeed, and observable only at very high speeds. For most practical purposes we can carry on with Newtonian ideas.

There are two morals which we can draw from the above story. The first is that the laws of physics may be correct, but within the limitations of our prevailing experimental ability. When newer and more precise observations are available, then the earlier laws could be challenged - they could have reached the limits of their validity.

The second lesson – though this is not so obvious – is that ultimately science is a unifier. Thus the Poincaré time dilatation or the Lorentz length contraction were both correct, though ad hoc. It was Einstein's work which at a single stroke brought out both these seemingly different results, it unified different concepts. In particular, we had thought of space and time as two separate unrelated concepts. Einstein's work showed that, rather we had to think of a single idea, spacetime. Our old concepts are however still approximately correct in the normal situations we encounter, for example the orbits of the planets around the sun and so on.

Einstein's two apparently contradictory postulates, together called the Special Theory of Relativity, no doubt explained the Michelson-Morley experiment. But is there a more direct way of experimentally observing, for example time dilatation? This time dilatation would mean that if your grandfather were in the Space Lab traveling very fast, with nearly the speed of light, then every hour by his clock would be several hours by your clock, so that ultimately, for example it is possible to think of a situation where a couple of years for him would be something like a hundred years for you. You could then end up being much older than your living grandfather when you greet him at the end of his trip.

Is there any way we can test this stunning, seemingly absurd prediction? Surprisingly the answer is yes. Traveling in a rocket with such high speeds is beyond our technological capability, but we could achieve the same conclusion in a more ingenious way. Neutrons it is well known decay into a proton and electron and a not fully understood practically massless particle called the neutrino, within a few minutes. We say that the half life of a neutron is about twelve minutes. This means if we have a hundred neutrons in a jar at this moment, after twelve minutes, there will be only fifty neutrons left, that is half of them would have decayed. However if the neutrons are moving very fast in a beam, what is called a neutron beam, it is found that their half life is prolonged, as if they live much longer. After twelve minutes, rather more than half the neutrons would be intact, that is would not have decayed. This is a dramatic confirmation of time dilatation. We can understand this strange, even commonsense defying idea that time dilates, by considering a simple clock. This consists of a source of light that can be switched on only for an instant and a mirror on top of the source. The light travels vertically upwards strikes the mirror and returns to be recorded by a counter. This interval is let us say one second. In one second light would have traveled a total of 300,000 kilometers, which means it traveled a 150,000 kilometers up and a 150,000 kilometers down in this time. That is like a round trip to the moon! But just suppose. Now let us suppose that we climb into our Space Lab with this clock as shown in the figure (Cf.Fig.2.4). For us, light would just go up and come down, but as seen by the stationary outside observer, light would have traveled along a triangular path. Each side is greater than 150,000 kilometers. So the interval between departure and return would be more than one second, if light were traveling at the same speed for both, as Einstein suggested.

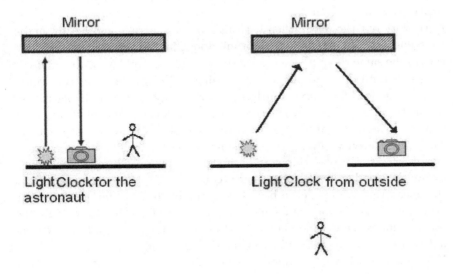

Fig. 2.4 Light Clock

This means that, firstly we or the astronaut would still calculate that one second has passed by this clock, whereas for the outside stationary observer, much more than a second has elapsed. So also for the neutrons in the beam, there is no change in their half life of twelve minutes whereas for the outside observer in the laboratory, the half life is much more than twelve minutes. We must remember that this is not just a physical effect. Everything – including the chemical, the biological, the physiological and even the psychological – is affected by the difference because time underwrites everything. In other words, your grandfather, stepping off the space ship at the end of the trip, would genuinely feel younger, and you would genuinely feel older!

In Newtonian mechanics there would have been no difference because the speed of light as seen by the outside observer would be a combination of the speed of our Space Lab and the speed of the clock. In other words it would be much more than the speed of light as noted by the observer with the clock. This extra speed would compensate for the extra distance traveled by light as seen by the stationary observer and the nett result would be that both observers would measure the same one second. So the crucial point here is that the speed of light is the same for all observers moving relatively at a constant speed.

If we work out the math, then it turns out that there is another dramatic consequence. You might even call it mindboggling. Suppose your friend sitting in the same room lights a cigarette, and at exactly that instant there is an accident on the road outside. According to Newtonian ideas this would be the case for all observers including, for example someone living in Andromeda galaxy. This is what we call the absoluteness of simultaneity – it expresses the fact that an instant of time has the same universal meaning. Not so in relativity! Two such events may appear very differently to different observers. For one, the lighting of the cigarette could take place much before the accident while to another it could be the other way round. Does this mean that a son can be born before the birth of his father? No. The order of events cannot change if one of these events is the cause while the other event is the effect. The order of events can change, only if they are not related by the cause and effect relationship.

Now you may ask what happens if the relative speed of two observers is not constant? Suppose one observer is moving with an increasing speed, that is accelerating, as seen by the other observer. Then the above simple logic would not hold. In fact Newton told us that acceleration is a manifestation of a force. So a new entity, force comes in with acceleration. This is the case with Newton's apple or any other object falling down to the earth. If we release that object from a great height, at that instant of release, the object is not moving, that is it has zero speed. As time passes however its speed keeps increasing, and by the time it hits the ground it would be moving much faster. For instance if the release was from the hundredth storey of a Manhattan building, its speed as measured from the fiftieth floor would be faster, but its speed when it reaches the ground would be even faster. This is of course due to the gravitational force, as we all know. The pile driver used in constructions, works on this principle.

Starting with this idea Einstein conceived of the following bizarre experiment, a thought experiment. A thought experiment is one which may not actually be performable, because of the technology or circumstances such an experiment requires. Nevertheless such an experiment does not violate any of the laws of physics, and would give a valid result if actually performed. To understand this experiment, let us again climb into our Space Lab, which we will use as an elevator – into outer space. Suppose you drop a ball from a height. This ball would immediately fall and hit the floor of the elevator. But now suppose that the Space Lab elevator were itself falling down with the same acceleration. This could happen in real life if the elevator cable snaps as it is coming down from the hundredth floor. Now what would happen is that the ball would be racing towards the elevator floor tugged by gravity, but the elevator floor itself will be racing away from it with the same speed and acceleration. So the ball just would not reach the elevator floor (Cf.Fig.2.5), (Cf.Fig.2.6). It is almost as if Newton's apple, having got detached from the branch of the tree remained there without hitting the ground! In other words it would be as if gravity had been switched off! Yes, the mighty force of gravity switched off!

Now we go further. Suppose you were in the Space Lab that is floating in deep outer space, far away from any stars, planets and so forth. You would then be floating weightlessly in the Space Lab. Now suppose by some means, for example using an attached high power rocket, the elevator were given an acceleration upwards of exactly the same magnitude as that encountered on the earth (but downwards). Then while you would have been floating alright, the elevator floor would come and crash into you just as on the earth you would fall and hit the elevator floor. In fact there is no difference between the two scenarios. You hitting the Space Lab floor, or the floor hitting you. Einstein argued that the force of gravitation could be removed or created by a suitable acceleration. This is a far cry from Newton's gravitation which was an actual force, much like the force with which we pull or push a wheel barrow for example.

In fact Einstein's proposal that gravitation and acceleration, to put it simply are equivalent solves another problem that was vexing him for a long time. This had to do with one of the experiments which made Galileo famous more than three centuries earlier. We all know how he was supposed to have puffed up the stairs of the leaning tower of Pisa and from the top dropped two spheres one made of wood and the other of metal. (According to some, Galileo did not actually do this!) [10]. Both the spheres are supposed to have hit the ground at the same time even though they had different weights. This demonstrated that the velocity and acceleration caused by the earth's gravity did not depend on the actual mass of the object concerned. On the other hand we also know that according to Newton's law of gravitation, this force depends very much on the mass. So some how the mass gets cancelled by his laws of mechanics: Because, the force is the mass times the acceleration.

We must pause for a moment to appreciate a nuance. There are two dif-

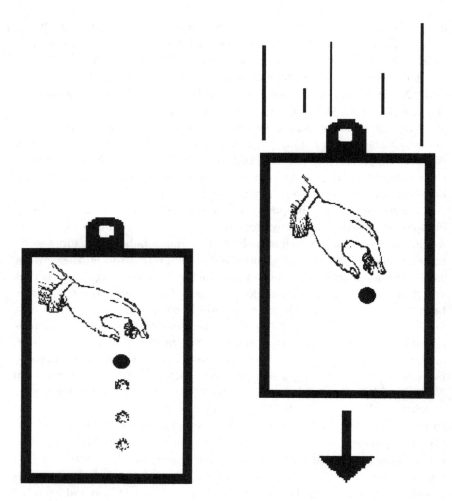

Fig. 2.5 Ball falling in a stationary elevator

Fig. 2.6 Ball floating in an accelerated elevator

ferent facets here. One is Newton's laws of mechanics which are true for gravitation as well as other forces. They are not gravitation specific. The other is, Newton's Law of Gravitation which has nothing to do with his laws of mechanics. In this law, he introduced a force, called the gravitational force and characterized it, quite independently of his mechanics. This force is directly proportional to the body's mass or "gravitational charge" much like the electric force is proportional to the electric charge. It so happens, almost miraculously, that the acceleration of mechanics is proportional to this same "gravitational charge". Not the electrical charge.

To get further insight, suppose we let go two spheres in our Space Lab, from a height. Further suppose that both these spheres are of very different sizes, material and masses. If the elevator were far far away, deep in inter stellar space, then like ourselves, these two spheres would float. Now the elevator begins to accelerate with exactly the same value as the acceleration due to gravity on the earth. In no time the floor of the elevator would rush and strike the two spheres at the same instant. (Or the two spheres would rush down and hit the floor at the same instant, looking at it another way). This collision would moreover take place at the same instant independent of the make up of the spheres. This is because the collision has nothing to do with the objects themselves, but rather is due to the movement of the Space Lab. Now Einstein was claiming that this motion which mimicked gravitation, was the same as gravitation. There was no separate force called gravitation. If this be so then, the acceleration of gravitation does not depend on the make up or mass of the material on which it is acting.

Next Einstein had to answer another question. He already had said that the laws of mechanics and electromagnetism were the same for all inertial observers including our astronaut rushing in the Space Lab with uniform speed. What about the accelerating Space Lab, which is no longer moving with uniform speed? Einstein claimed that the astronaut would also deduce the same laws of mechanics and electromagnetism as the inertial observer, including an astronaut in an uniformly moving spaceship. We saw that the equivalence of the laws of mechanics and electromagnetism for uniformly traveling astronauts could only be possible if we profoundly changed our ideas of space and time.

It turns out that claiming these laws to be the same for accelerating astronauts too, would lead to some equally profound implications. A simple way of seeing what this means in electromagnetism is to consider the motion of a ray of light, which we know always moves along a straight line–even with special relativity. Let us now consider a ray of light that is flashed diagonally across our Space Lab.

Suppose first, the spaceship is at rest, that is floating freely in deep outer space. A ray of light is flashed from one side of the elevator to the other, parallel to the floor–or so says our astronaut. At that very instant the elevator starts moving downwards with an uniform speed. Our description from outside would be slightly different. By the time the ray of light traveled one centimeter, parallel to the floor, the floor itself has come down by let us say one millimeter. By the time the ray would travel another centimeter, the uniformly moving elevator would come down another millimeter, carrying the floor with it and so on. The nett result is that the ray of light would to us who are outside appear to move at an angle to the floor, but nevertheless in a straight line (Cf.Fig.2.7).

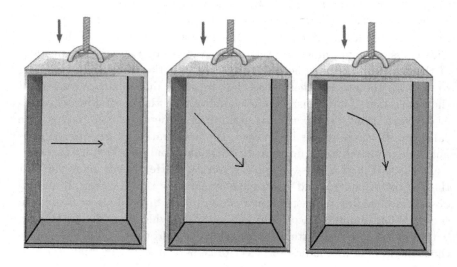

Fig. 2.7   The deviation and bending of light

Suddenly the spaceship is accelerated. Let us say that as before, by the time the ray of light travels one centimeter parallel to the floor our Space Lab comes down by one millimeter. But by the time the ray of light would have traversed the next centimeter, the floor would have come down by more than one millimeter because its speed is now no longer constant but rather is increasing. Let us say it has moved by two millimeters. By the time the ray of light travels yet another centimeter, this time the floor would have plunged down by rather more than two millimeters, let us say by three millimeters. If we now trace the path of a light ray as seen by us from outside, it would not only appear slanting, but it would actually be a curve. That is light, contrary to everything that we have been taught appears to be bending. Light would no longer appear to travel along straight lines! (Cf.Fig.2.7) This is the new effect which we will observe if we accept Einstein's claim that all laws of physics, including electromagnetism would be the same for all observers, irrespective of whether they are moving uniformly or accelerating.

One can put forward any consistent hypothesis or model, but ultimately whether this model is meaningful or not depends on the predictions it makes and whether these predictions can be tested. Clearly no one would accept Einstein's bizarre ideas without solid observational proof. This came about just a few years after he predicted what may be called his gravitational bending of light. It is called gravitational bending, because as we saw acceleration and gravitation are one and the same.

In 1919 there was a total solar eclipse which was visible from parts of Africa. Such an eclipse could provide an opportunity to test Einstein's ideas. This is because when the sun gets eclipsed, that is its face is just covered by the moon which comes exactly in-between at that time, then we begin to see the stars which are pretty close to the Sun. Close here means, close as they appear. Of course the stars are there even if there were no eclipse, but then we cannot see them. So suppose we observe a nearby star just as it becomes visible when the eclipse starts. This means light from the star is reaching us. This light has to graze past the sun, because the star is actually much much farther away. If this ray of star light should be bent by the sun's gravitation, that would mean that the star as we see it would appear a little higher than it actually is.

Sir Arthur Eddington the celebrated British Physicist headed an expedition to the West coast of Africa to check this out and photographed stars very close to the Sun. He then compared their positions on the photograph to their actual catalogued positions and found that indeed they appeared in the photographic plates, that is to our eyes as being a little shifted upwards. This was a dramatic confirmation of the strange theory of gravitational bending of light, (Cf.Fig.2.8).

There are other forces which can be "created" by motion and which are coupled to the gravitational "charge" or mass, exactly like the force created by the accelerating spaceship. For example let our Space Lab, which till then was either uniformly moving or stationary, suddenly get into a spin. The astronaut would be pushed to the side of the spaceship due to the centrifugal force. This is the same force that you feel on your hand, when you hold the open door of your car as it swerves - the centrifugal force would throw the door open if you were not pulling it in. This centrifugal

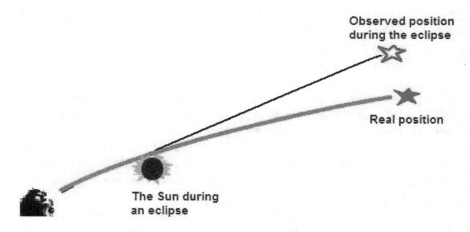

Fig. 2.8   The bending of light observed during a total solar eclipse

force depends on mass in the same way as gravitation or the force of acceleration - if the astronaut were twice as massive, he would feel twice the force. There is no reason why we cannot carry over the above reasoning of the accelerating spaceship to this spinning spaceship. Once again we can conclude that a ray of light would bend towards the direction in which the force is pointing.

What does this bending mean? Suppose light were like a boy running through a narrow corridor If the corridor were circular, then the boy concentrating on the ground below would think that he is running straight. He would be very surprised to find himself back where he started after running the length of the corridor in what he would suppose is a straight line. Einstein said, so also the ray of light would be traveling in a "straight line" through space, except that this space or more correctly spacetime is bent like the corridor. The bending of spacetime was yet another effect that Einstein introduced - this time it was the effect of gravitation or the equivalent force due to acceleration.

We already saw that time gets dilated because of speed - the greater the speed, the greater the dilation. Remember the fast moving neutrons and their life time which increases with speed? Now let's see what happens if our Space Lab were crashing on to the earth. As it approaches closer to the earth, under gravity, its speed would increase. So at a certain height time would have dilated by a certain amount. That is, one second for the astronaut would appear to be two seconds on the earth for us let us say. Soon the astronaut would be closer to us and would be crashing even faster. So his one second would be even more to us, shall we say it is four seconds. This means at a certain height the one second is two seconds and at a lower height it is four seconds, all this being due to gravity. There would thus be a difference in time intervals at different heights.

There is another way of looking at this. Let us consider, for example the sun and the earth. As light leaves the sun, it begins to loose energy, because the gravitational field of the sun drops with distance. Loosing energy means that the frequency of this light begins to fall, because as we will see, the frequency or rapidity of oscillation is directly related to the energy. So if we are on the earth, it would appear, using our light clock again, that time intervals near the sun are longer. Equivalently time slows down near the sun or speeds up near the earth which is much less massive compared to the sun, and so has less gravitational energy.

To quote the inimitable American physicist, George Gamow [11], "A typist working on the first floor of the Empire State buildings will age more slowly than her twin sister working on the top floor. The difference will be very small however; it can be calculated that in ten years the girl on the first floor will be a few millionths of a second younger than her twin on the top floor."

We must remember that when we say time slows down, that means everything that goes with time also slows down - the rates at which electrons orbit atoms, the rate of chemical and biochemical reactions and so on as already noted (Cf.Fig.(2.9)). So with a suitable arrangement, we could ensure, in principle, that one of the twins is an old hag while the other is still a babe in arms! Space and time are much more mysterious than we can ever imagine!

You might well ask, "We could verify that time dilates in special relativity, for example by observing neutrons in a beam. Can we similarly check out this gravitational time increase?" Not so likely with the Empire State Building, as you might guess. Nevertheless experiments have been performed, starting 1962. Two very accurate clocks were used, one at the top and the other at the bottom of a water tower. The clock below, being nearer to the earth was in a stronger gravitational field, and indeed it was found to run slower, in agreement with general relativity.

Moreover we must remember we are constantly receiving powerful radiation from massive objects in deep outer space, objects like quasars. Because of the power gravitational fields at these objects, compared to the low grav-

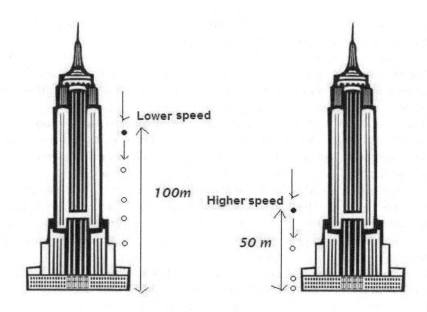

Fig. 2.9   Different speeds of a falling object at different heights

itation on the earth, by the time the radiation reaches us, it would have traveled through different "time zones". Time zones in the sense that intervals there would be different: seconds there compared to seconds here. This would effect the frequency or wavelength of the radiation. It is called the gravitational red shift, because light from there wouldbe redder than its counterpart light here. It is believed that such gravitational red shifts have been observed, though by no means is this verdict unanimous.

Chapter 3

God does not play dice, or does He?

"It may happen, it may not happen"

– Ancient Indian saying.

3.1 The Classical Catastrophe

Classical physics developed over the 17th, 18th and 19th centuries to reach a degree of precision and perfection that was as impressive as the Ptolemaic model, two thousand years earlier. Indeed, delivering the inaugural lecture at Cambridge University in 1871, the doyen of electromagnetic theory, James Clerk Maxwell observed "In a few years, all the great physical constants will have been approximately estimated, and ... the only occupation which will then be left to the men of science will be to carry these measurements to another place of decimals." [12] Classical physics later came to include also Einstein's special and general theories of relativity.
So how does man's concept and knowledge of the universe develop? We could well imagine that it develops progressively, one step leading to another. But then the Greek Ptolemaic model held sway for some fifteen hundred years and then collapsed in toto. Thomas Kuhn, the philosopher of science suggests that this is how science progresses [13]. It develops impressively up to a point and then collapses unable to confront new observations, so that a new edifice based on new thinking is built up over the ruins of the past. This is much like the monuments left behind by history. However difficult it was for a scientist to have imagined in the earlier part of the twentieth century, classical physics too suffered a similar tumble. It was a beautiful theory that explained just about everything–but in the large scale, the macro or gross world.

New experimental inputs came in from an unexpected frontier – a study of atomic systems in the early years of the twentieth century. These threw up results which just could not be explained by classical physics. They formed a huge enigma - it was only by discarding the old concepts and invoking radically new hypotheses that the new evidence could be explained. These new experimental findings came from different directions - from optics, from thermodynamics, from atomic physics and also from completely new fields [14].

In optics for example, it was well known that light was emitted by substances when they were heated. For instance if an iron rod is inserted into a burning flame, after a certain time, it begins to glow. We could analyze the light which the rod emits by looking at it through an instrument called the spectroscope which, to put it simply, breaks up light into its constituent colours by using a prism. We then begin to see colored lines. Classical physics could explain why the rod should emit light - it would say that the heat of the flame causes the electrons in the rod to accelerate, and when electrons accelerate they emit light or more generally electromagnetic radiation.

However what puzzled scientists was that the experiments brought out a series of such spectral or colored lines and not a rainbow type of a continuous spectrum. For instance, when a Hydrogen atom is heated, it emits a characteristic set of such glowing lines - this set is called the Balmer series of lines. This series of lines was anything but random. Physicists could experimentally find a set of what are called spectroscopic terms that could be associated with the atoms of a substance. These mysterious numbers had an interesting characteristic - their differences gave the wavelengths or what is the same thing, the frequencies (which are reciprocal of the wavelengths) of the color of the spectral lines. This was called the Ritz combination principle, after its discoverer. The puzzle was, firstly the discrete lines and wavelengths rather than a continuous spread and secondly why this mysterious prescription called the Ritz combination principle put forward in 1908? There was no way that classical physics could explain this.

Another crisis for classical physics came from thermodynamics. It was well known that atomic systems could absorb or emit energy, when in contact with a suitable environment. As in the case of the light spectra, it was thought that this took place continuously. To put it a little more technically, the spectrum of radiation emitted by a hot object is continuous much like the continuity in colors we see in a rainbow. Or, so it was thought. We also knew that this spectrum covers a wide range of wavelengths or frequencies. Moreover there was a wavelength at which there is a maximum intensity of radiation, and this depended on the temperature of the object that was radiating. This was the famous Rayleigh Jeans law, named after two famous British physicists, Lord Rayleigh and Sir James Jeans. According to this "law" the energy per unit volume is proportional to the square of the frequency (or inversely proportional to the square of the wavelength). This law agreed well with experiments - but as was realized, only at low frequencies or large wavelengths. The problem is that if we follow this law, then as the frequency increases the energy too increases without limit, which of course is impossible. This was called the ultra violet catastrophe because ultra violet rays have higher frequency compared to heat or light. It was an impasse for classical physics.

Then there was another interesting experiment which brought out what is called the photoelectric effect. This is crucial to us today because many modern optical devices are a direct consequence. These experiments show that when electromagnetic radiation strikes certain metals, then a stream of electrons is ejected. This stream constitutes what is called photoelectricity. To put it roughly, the puzzle here was that the energy of the emitted electrons depended on the frequency or wavelength of the incident radiation and not its intensity - the greater the frequency, the faster the electrons moved. By the way, an undulating wave has two lengths associated with it. Its height and width. The height measures the intensity, like the volume of sound. The width on the other hand describes the wavelength or frequency, like the pitch of sound. All this was again inexplicable from a classical point of view.

Historically the escape route from the ultra violet catastrophe of black body radiation, came first. This was due to the pioneering and ground breaking work of the German physicist Max Planck who in December 1900 conjectured that energy was not continuous but rather came in discrete packets or quanta. Furthermore the energy content of these quanta depended directly on their frequency. Once these non continuous and discrete quanta are introduced into the Rayleigh Jeans formula and subsequent calculations, the total energy turns out to be finite, as it should be. Revolutionary and even absurd as the idea might have been at that time, it had to be admitted that there was no other way out of the ultra violet catastrophe of infinite energy. In fact Planck himself distanced himself from this radical idea for many years [15].

Later, in 1905 Einstein used the new and radical theory of Max Planck to explain the photoelectric effect. He argued that the radiation impacting the metal plates behaved not like waves, but rather like the discrete packets or quanta of Max Planck. This particle behavior of supposedly continuous radiation could then explain how the electrons could be ejected by the metals as a result of the impact. In one sense, as we saw, Planck's idea was not as new as it might seem. We know that, while Christian Huygens in the seventeenth century believed that light was a form of a wave, Newton had suggested that it was made up of corpuscles.

Around the same time Geiger of the Geiger counter fame and Marsden were performing experiments of another kind. Streams of what are called alpha particles which are nothing but the nuclei of Helium atoms, bombarded a thin gold foil. They passed through the foil and were detected on the other side. Classical ideas told us very unambiguously what was to be expected: A random set of points indicating the places where the alpha particles struck the detector. The actual result however was anything but this. Some of the particles had come out without much deviation as if they sailed through empty space. But a few – one in about twenty thousand had deviated quite a bit. There was'nt the random expected collection (Cf.Fig.3.1). How could we explain this discrepancy?

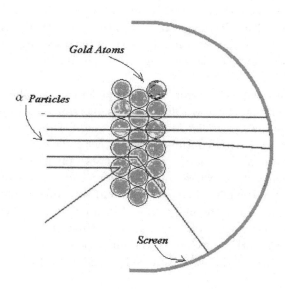

Fig. 3.1   Rutherford's Gold Foil Experiment

Till then it had been supposed that all matter was made up of electrons and perhaps protons in a sort of a random mixture. But the experiments could only mean that matter, that is atoms were very different. Sure, there were electrons. But these were far away from heavier central nuclei, with nothing but empty space in between. It's a bit like the solar system with the sun at the centre and then the planets spaced out at different distances, without anything in the intervening space. So those particles which went through the empty spaces came out without much of an effort, while those which encountered the central nucleus were knocked about quite a bit. There was hardly any third category.

So, could this mean that the atom was made up of a heavy central part, the nucleus with electrons relatively far away? This was proposed by an experimental physicist of New Zealand origin, working at the Cavendish Laboratory in England, Ernest Rutherford in 1911. But this leads to trouble. Because it was well known in classical theory that a collection of charges could not remain in equilibrium, if these charges were stationary. What would happen in this case is that the slightest of disturbances would send all the particles crashing into each other, and such an atom would be destroyed in less than a trice! In a sense, the universe of today would not exist!

This could only then mean, that the lighter electrons were not stationary, but rather like our planets, they would be orbiting the central nuclei. The atom began to look more and more like our solar system. Except that classical theory did not allow even this. That is because when an electron rushes through an orbit, it radiates electromagnetic waves, thereby loosing energy. Such an orbiting electron would as a result spiral in towards the central nucleus. It is much like some of the space crafts which have spiraled in to crash into the earth, as they lost their speed due to frictional contact with the upper atmosphere. This way too, the atom would be destroyed in no time. Either way the universe would cease to exist in less than a wink. Some impasse! There appeared to be no way out.

A Danish Physicist, Neils Bohr joined Rutherford at Manchester. He put forward in 1913 a few bold ideas to resolve the impasse much as Einstein had done a little earlier. Bohr proposed a set of "illogical" postulates, that is 'don't ask me why' type of ideas. Their merit lay solely in the fact that they showed a way out of the impasse. Thus he proposed that electrons could continue to orbit the heavy central nuclei, but without loosing any energy. This of course blatantly contradicted the laws of classical physics. Bohr claimed that electrons could use different discrete orbits, each with its own energy - but here the analogy with solar system planets breaks down. A planet in the solar system is free to orbit the Sun with which ever energy is given to it. Bohr on the other hand restricted the energies only to certain allowed energy levels, that is a series of allowed values, and no value inbetween any two neighboring values. It is a bit like saying that the electron would orbit the nucleus with the energy one or the energy two or three and so on in some units, ignoring the infinite number of values between one and two, or two and three and so on.

Bohr went on to say that an electron with a certain energy could jump to an orbit with a different energy that was greater or even smaller. If the electron jumped to a smaller or lower energy orbit, then the difference of energy would precisely be some multiple of Planck's quantum, for example it could be two quanta or three quanta and so on. Conversely if the electron in jumping to another orbit gained energy, then it would have absorbed for example two or three or what ever number of quanta of energy from outside.

Bohr related the number of quanta of energy that an electron would emit with the spectroscopic numbers which gave if you remember, the spectral lines of the Ritz Combination Principle - one of the first of the paradoxical findings. Thus at one stroke Bohr could explain several experiments, although with the help of rather arbitrary and against the prevailing physics commonsense assumptions. However, Bohr hesitated to publish these radical ideas. Would anyone take him seriously? After he returned to Copenhagen, he was persuaded by Rutherford to go ahead and do so.

Albert Einstein too invoked Planck's quanta to explain the photoelectric effect. Thus, he argued, if the electromagnetic radiation impacting the metal plate were a series of quanta, rather than a stream of waves, then these quanta could kick out electrons which lay trapped in the middle requiring this exact same amount of energy to be liberated. That is how, the energy of these quanta or equivalently the frequency or wavelength of the radiation is crucial: waves with the wrong quanta, though plentiful in number would not do the trick. They would be like square pegs in a round hole.

These bold ideas constituted what we today call the Old Quantum Theory. It was both radical and ad hoc. It matched the junking of the Greek heavenly crystal spheres, acceptable only because this was the only way to explain experiments whereas the older ideas could not. What all this meant was that, in a sense, we had returned to Newton's earlier concept of light as being a stream of corpuscles, rather than a stream of waves as supposed by Huygens. It is rather curious that Albert Einstein got his Nobel Prize for explaining the photoelectric effect, rather than for his deep studies of space and time. This was because, most scholars could not clearly understand Einstein's spacetime ideas [16].

3.2   Particles Disguised as Waves

So it was enough to consider electromagnetic radiation to be made up of discontinuous quanta or packets of energy just as we had convinced ourselves that these were continuous waves. But what followed challenged our thinking even more. These iconoclastic results were summarized by De Broglie a French nobleman of Italian origin, in his PhD thesis. He suggested in 1924 that matter too behaved in a contra manner: like waves, rather like the discrete standing waves on a string plugged at both ends, a guitar string, for example (Cf.Fig.3.2). And why not, if waves could behave like particles? The PhD committee was much too puzzled to pass a judgement. The members consulted Einstein who approved of this idea. Only then did De Broglie get his degree!

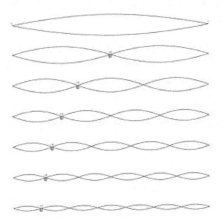

Fig. 3.2   Standing Wave

However, soon experimental confirmation for this strange idea came by. These were experiments performed in 1927 by two American physicists, C.J. Davisson and L.H. Germer. They bombarded an intermediate screen with a stream of electrons, which then came out on the other side. Again classical physics was quite clear about what to expect at the detector. It would be like a random peppering of electrons. This is what happens, for example if we shoot at a target with an automatic rifle - the bullets would be scattered randomly in a region whose size of course depends on the expertise of the shooter. What actually was observed was anything but that. There were alternately regions with no electrons and with a concentration of electrons. This is very familiar from experiments with light - we get what are called Newton's rings - alternately dark and bright regions. But then light was a sort of a wave and here the samething was happening with electrons which are clearly particles. Puzzling to say the very least. But further proof followed the same year.

G.F. Thompson bombarded a beam of electrons on to a thin gold foil. Again he got on the other side the same alternate rings. It is easy to explain why there should be these shadow and bright regions, in the case of light. Different waves of light would interfere with each other. Remember a wave has a high and a low. Where highs and lows meet, they can destroy each other. Where a high and a high meet they can reinforce each other. So we have dark regions with waves that have destroyed each other and bright regions where they have reinforced each other (Cf.Fig.3.3), (Cf.Fig.3.4).

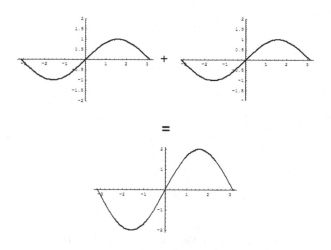

Fig. 3.3   The addition of waves

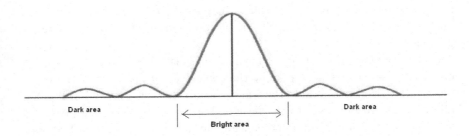

Fig. 3.4   Interference of waves

Experiments were screaming out to us that even electrons behaved the same way. De Broglie had been vindicated. The whole beautiful structure of classical physics was collapsing, much like the wonderful crystal spheres that the Greeks had sculpted. Now anything was possible.

Other experiments too were performed which brought out different facets of the new school of thought. For instance the British physicist A.H. Compton in 1923 was scattering X-rays through light elements. These experiments too showed that X-rays, which were electromagnetic waves, behaved once again like quanta or packets or photons.

There was here a duality - particles behaving like waves and waves behaving like particles. As the famous British physicist, Sir W.H. Bragg would put it, "Physicists use the wave theory on Mondays, Wednesdays and Fridays, and particle theory on Tuesdays, Thursdays and Saturdays." (Mercifully, Sundays are off days!). All this needed a new mathematical description, because particles were known to obey Newton's laws of mechanics while waves obeyed Maxwell's theory of electromagnetism. Now particles too had to be included in a wave type of description. This was the beginning of the second or new phase of Quantum theory.

In this new phase, particles were described by a wave equation. This pioneering work was sphereheaded by the Austrian physicist, Erwin Schrodinger, and the equation bears his name. It was the result of a question put to him when he was speaking about De Broglie's matter waves. What is then the wave equation, he was asked: Schrodinger's answer was his equation. There was however a lacuna - it was very well to describe a particle as a wave, but what does this mean in real terms? In the case of particles we know that the particle is here or there or elsewhere. A wave on the other hand is spread all over. So how can a wave be describing a particle?

The answer was supplied by a German physicist, Max Born who used a commonsense approach. We know that where the wave is high, there the intensity is also high. Where the wave is small there is a low intensity. We all know this from a visit to the beach. A high wave blows us away but a low wave is still pretty calm and we can continue to stand in it. Max Born connected the waves associated with particles to this everyday experience - where the wave of the particles is large in magnitude he suggested that there is a greater likelihood of finding a particle than where the associated wave has a small magnitude.

As Born himself put it [17], "Again an idea of Einstein's gave me the lead. He had tried to make the duality of particles – light quanta or photons – and waves comprehensible by interpreting the square of the optical wave amplitudes as probability density for the occurrence of photons. This concept could at once be carried over to the (Schrodinger equation for the wave function)."

Max Born got the Nobel Prize for this ingenious interpretation, but almost missed it! This is because in part of the paper he did not relate the magnitude of the wave (this measured by the square of the wave function) to the probability of finding the particle. He merely took a wave itself. Later in the same paper in a footnote he suggested his correct interpretation.

The idea that a particle is a wave explains a mysterious prescription given by another doyen of those heady times, the German Werner Heisenberg. Essentially his Uncertainty Principle proposed in his doctoral thesis said that you cannot precisely measure the position of a particle (and therefore also its speed or momentum). How can you, because the particle is a smeared wave rather than a precise point, unlike in classical theory (Cf.Fig.3.5).

Heisenberg devised an experiment, which we can do in principle in our Space Lab. Let us try to measure the position and velocity of a moving electron. To do this, we need to flash a beam of light or any other electromagnetic radiation on to the electron. The radiation (or photon) then bounces back to us and carries the required information with it. But in the process, the photon has disturbed the electron unpredictably – both its position and speed have been changed by the collision. So what was the electron's position or velocity? This is where classical or Newtonian physics diverges from Quantum theory. As we deal with macro or large objects classically, the impinging photons do not disturb them and bring back

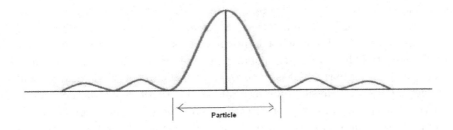

Fig. 3.5   The uncertainty spread of a particle

accurate information. In any case, if classical physics resembles a ramp, Quantum theory is like a staircase.

There was still some work to be done however. This new Quantum theory was based on waves, but did not take into account Einstein's relativity. The British-French physicist P.A.M. Dirac achieved this [18]. He formulated an equation that combined both the wave Quantum mechanics and the theory of relativity. Some strange and stunning results tumbled out of this new marriage, the Dirac equation. The first was that an electron would have an angular momentum, rather like a spinning top. But the value of the electron spin as it is called cannot be deduced from the theory of spinning tops. To put it roughly, a top can spin with its axis in any direction. For example the earth spins on its axis that makes an angle of twenty three and a half degrees with its plane of orbit. This angle is different for different planets, and even for the earth it could have been anything else. An extreme case is that of Uranus which spins in a plane practically perpendicular to its orbital plane. The spin of the electron however not only had a different value from what classical mechanics would give us, but also this value could be only up or down. The electron could not spin at any intermediate angle. Its axis of spin would be either up or down. The other strange result that tumbled out of the Dirac equation was that like the electron, there would be what you may call a mirror particle, the positron which differed only in its charge. Such a particle is also called an anti-particle.

So much for theory. Could we actually experimentally verify these strange conclusions that seem to make no sense - certainly not from the point of view of classical physics. The answer soon came. Yes. The experiment to test the spin prediction was simple enough. If a charged particle like the electron spins, it is like a small electrical circuit, rather like a current moving through a coil. We know what happens in this case. There is a magnetic field generated that is perpendicular to the plane of spin. So if a beam of charged spinning particles passes through a strong magnetic field, what will we get? Classical physics tells us that the magnetic fields of the various particles would be in different directions and the external magnetic fields would deflect these particles into different directions. In other words once again we should get our random peppering of the particles on a detector. Quantum mechanics on the contrary tells us that the magnetic fields of the particles would be along two directions only, up that is along the external magnetic field, or down that is anti parallel to it. Then we should get just two regions with particles – one on top and one below (Cf.Fig.3.6). That is exactly what the experiments performed by Otto Stern and Walter Gerlach show. (This experiment was actually performed some years earlier with diverse interpretations). Similarly Dirac's strange beast, the anti-particle was also experimentally observed.

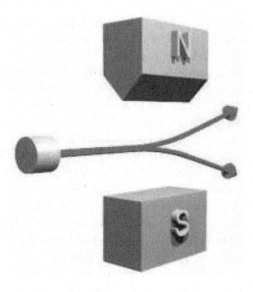

Fig. 3.6   The Stern-Gerlach Experiment

The old and new Quantum theories brought out many strange and commonsense defying concepts. We could even say they were impossible concepts. Yet the experiments all bore out these ideas. Professor Abdus Salam the Nobel Laureate would say, "Experiment is at the heart of Physics" as noted. However reasonable and persuasive may be the arguments put forth by commonsense and however absurd other arguments may sound, ultimately experiment is the arbitrator. There are, as we will see, any number of conceptual issues that Quantum mechanics has thrown up which are still being debated inconclusively more than a century later. But of this there is no doubt. Its rules explain nature, unlike the earlier ideas.

Moreover there is no doubt that Quantum mechanics has proved to be not just a spectacular and successful theory, but an intellectual gift of the twentieth century to mankind. Whether it be in science or in technology, we hardly realize and recognize the deep role that Quantum mechanics plays in our daily lives. For example it is indirectly behind the nuclear power which we use today. This was possible because it led to an understanding of radio activity itself. It is also the theory behind semi conductors and super conductors. It has led to the invention of the maser and laser and also to the interaction of radio waves and nuclei which is behind Magnetic Resonance Imaging or MRI that has become so important these days in medicine. The list can go on and on.

Apart from such technological benefits, Quantum mechanics has made many predictions which have been actually verified in real life. For instance Dirac's theory of the electron predicted the existence of anti matter. Anti matter was subsequently seen–anti protons, anti electrons or what we call positrons all of them have been identified. Matter antimatter collisions create huge amounts of energy. Who knows? This may be a future source of energy. We will briefly return to some of this later.

Perhaps the peak of Quantum mechanics in the scientific sense has been the formulation of what is called Quantum Field Theory or QFT for short. The point is that in Quantum mechanics we are essentially dealing with what may be called isolated particles. But the real life scenario is that of any number of particles. Such a large number of particles can be understood, and we can carry out calculations in terms of what is called a field theory. We must bear in mind that even in the realm of Newtonian mechanics we cannot really solve even the three body problem as the great French mathematician Poincaré showed us more than a hundred years ago [19]. So what about the universe which consists of millions of particles, billions of them? We treat all the particles as forming a field and then devise special techniques for dealing with the field as a whole. We can pinpoint a particle much like we can, points on a sheet of paper. There are problems in this approach, of course. For instance we encounter infinite quantities which are meaningless in the real world. Clever mathematical techniques are invoked to bypass this problem. QFT has proved to be enormously successful, though the underlying math leaves many uncomfortable. Perhaps it is more like an algorithm that works.

QFT had led to many beautiful explanations for Quantum phenomena. It has also led to a comprehensive theory of not just electromagnetic interactions but also other interactions we had to contend with in real life such as what are called strong interactions.

What are these strong interactions? Consider the nucleus of an atom. This consists of some protons all of which have positive charge and also there are a few neutrons. The question which you might ask legitimately is, how can all the protons stay together? After all they must be repelling each other. The answer is that, yes protons repel each other provided they are not too close. There is another force between protons which is called the strong interaction or strong force which has a very short range. That is, it operates only when protons are very very close. This force is attractive as if suddenly protons, have acquired opposite charge except that this force is much stronger than the repulsion of the protons due to the electric charge. So this is one force apart from the electromagnetic force. Then there is what is called the weak interaction.

This force causes a neutron to decay. Remember neutrons have a half life of about twelve minutes. In that time a neutron tends to break out as a proton, an electron and a mysterious particle, a not very well understood particle called the neutrino. A very strange particle because it defies the left right symmetry which otherwise seems to rule in real life. Such particles were predicted by Wolfgang Pauli way back in the 1930s and subsequently were identified and detected. This entire process of the break up of a neutron is caused by an interaction or a force which we call the weak force which is much weaker than the electromagnetic force.

Quantum Field Theory has been able to explain or give a reasonably satisfactory framework for all these interactions. That does not mean that it has solved all the problems. There are still a number of questions which Quantum theory and QFT have to answer and we will look at some of these problems a little later.

The point is that perhaps no other theory has been so successful in human history as Quantum theory. However Quantum theory has many strange aspects which conceptually challenge and defy the human brain. Remember so far our approach to Quantum theory or the problems of physics has been pretty utilitarian. We have tried to explain experiments and in the process we have in fact made some predictions to be verified by further experiments. That has been the way of science. But you may wonder that all this is like a mechanical fix, like a mechanic who comes and repairs a broken down equipment, a car for example. There are many concepts that are yet to be understood in clear terms and therefore concepts about which a debate still rages on.

This debate was launched by two of the greatest intellects which the human race has produced: Einstein and Bohr. So it's really something very deep that we have not yet been able to fathom.

Let us start with an experiment put forward by the brilliant American physicist Richard Feynman [20]. According to him this experiment captures the heart of Quantum mechanics. It is called the Double Slit experiment. So let's enter our Space Lab again and set up a screen with two thin slits or openings side by side. There is also a detector behind the screen. We bombard the screen, with let us say a stream of electrons. Then let's watch what happens on the other side.

From a classical point of view an electron would go either through slit 1 or slit 2 and come out at the detector. What we would see there is a random series of dots as we noted before. However if electrons are waves then the story is different. Now what happens if waves or electrons pass through both the slits? We expect that these two waves will interfere with each other. It will be a bit like the continuation of the single slit experiment. But wait. What we observe at the detector is this: Electrons still come like bullets in the sense that it is not that we are getting half an electron. In that sense electrons arrive like particles and are detected by the click of a detector which could be a Geiger counter so that whenever a single particle arrives there is a click. If the particle doesn't arrive there is no click. Nothing inbetween. But if we look at the pattern of the electrons which arrived at the detector or the screen to which the detector is attached, we find that it is more like two waves passing through the slit no 1 and the slit no 2. More like light waves with interference patterns unlike in the case of bullets in a classical experiment.

So here we have indeed a strange situation. The electron behaves like a particle and it also behaves like a wave *in the same experiment*. This is all very mysterious. Moreover we cannot say that the electrons are coming either through slit number 1 or slit number 2. If that had been the case then we could have carried out two experiments separately. First we could cover slit number 1 and allow the electrons only to go through slit number 2. Next we could do the opposite, that is cover slit 2 and allow electrons to go through slit number.1 As electrons are behaving like bullets or particles we would expect that when both slits are open what we get at the detector is the sum of the two.

But that is not the case. We get something that is neither through slit number 1 nor slit number2. That is the wave nature. We cannot say that an electron is going either through 1 or through 2 something which we can say with bullets. As Richard Feynman put it, "in reality it (this experiment) contains the only mystery. We cannot make the mystery go away by explaining how it works. We will just tell you how it works. In telling you how it works we will have told you about the basic peculiarities of all of Quantum Mechanics." (Cf.Fig.3.7).

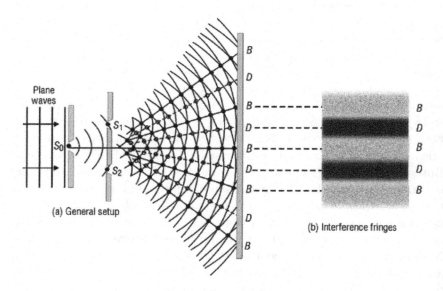

Fig. 3.7   The double slit experiment

We can now perform the same experiment with two slits but this time with a variation. Suppose somehow we use a laser beam or whatever to observe the electrons as they pass through the slits. That way we will know through which slit the electron has gone and we will be able then to say that the electron has gone through slit 1 or through slit 2. Will it then be like bullets and classical physics where too we have such a knowledge? The answer is, yes! When we observe the electrons the mysterious interference disappears. By the very act of observation we have reduced the mysterious Quantum mechanical experiment to a classical experiment. Here observation means an observation of the electrons to know through which slit they have passed.

What we can conclude from this latter experiment is that the Quantum mechanical wave function no doubt contains both the possibilities of the electron going through slit number one or the electron going through slit number two. So the wave function is a mixture of these two possibilities. However when we observe the electron, in other words when we bombard the electrons with a laser beam for example, then the wave function "collapses" into one or the other of the possibilities. After we observe, it is as if we are back, in classical physics.

The important moral of this experiment is that the act of observation influences the outcome. Well, let's analyze this in a little greater detail.

The wave function really represents a wave and as we have seen waves can interfere: You can add waves or superpose them. So a wave is in a sense a combination of more elementary waves which represent elementary or actually measurable possibilities. The final wave is therefore a combination of different possibilities. However when we observe a wave what happens is that some form of radiation (or photons) goes and strikes the wave and returns to us to give us the information. This act of striking the combined wave with radiation causes the combined wave to collapse into one of its elementary waves which represents a single possibility. Unfortunately though we cannot predict in advance, which possibility will be realized. So that explains the above mysterious experiments. When we observe the electron to determine the exact slit through which it passes, we destroy its composite wave structure and it collapses into a classical particle. There is no escape from this. It is a bit like a school boy whose behavior depends on whether the teacher is watching or not.

3.3 The Paradox of Reality

Einstein however could never reconcile himself to this type of an argument. For him nature has a reality, an identity that transcended these measurements. It is not a mixture of possibilities. Nature is either this or that, irrespective of whether you are observing or not observing. It's a bit like saying the moon is there whether you observe it or not, whether there is life on the earth or no life on the earth. An objective reality as you may call it exists in nature. So believed Einstein. But what the experiments tell us is quite a different story. The so called actual reality if we may call it that, becomes a reality only when we observe it. Till such a time there is no actual reality or objective reality of the type Einstein believed.

Now this was something which troubled not just Einstein but also the other founding fathers of Quantum theory. Schrodinger for example thought up another experiment which is generally called Schrodinger's cat experiment. This is very bizarre. It consists of a cage inside which there is a cat. It could be a parrot or a mouse or any other creature of course. The cage is covered. But the cage also conceals a deadly secret, a radio active element which can release lethal radiation capable of killing the cat. However the element is in a state of "release" and "no release"– it may release or it may not release the lethal element. It's really a combination of the two possibilities according to the strange Quantum logic. Only when we open the cage or the cover of the cage and light goes in, for example, by flashing a torch into an otherwise dark cage, which then strikes the element and either makes it radiate the lethal radiation or makes it not do so that we choose though inadvertently, one of the two possibilities.

Now what about the cat before an observation is made? If we follow our Quantum logic through all the way, the cat is neither alive nor dead. These are the two possibilities. It could be one or the other but which one? This will be realized or will be made real only by an observation. So before you observe you have the cat that is neither alive nor dead, somewhere in between. Poor cat! When we open the cage and allow light to flood in, the light strikes the radio active element and there could be lethal radiation as a result in which case instantly this radiation will kill the cat and what we will see is a dead cat. On the other hand it is equally likely that the light which we flash into the cage puts the radio active element into the second state where it does not radiate the radiation. Then what we will find is a cat that is alive. (Cf.Fig.3.8) What a bizarre experiment! We don't know if the cat is alive or dead. We don't know if the moon is out there or not till an observer on the earth actually observes and finds the moon or finds that there is no moon. In effect this is what it means. There is no objective reality or a reality out there, independent of our observation.

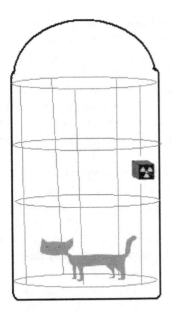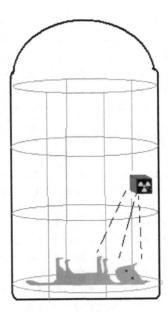

Fig. 3.8 Schrodinger's Cat

An observation therefore triggers the collapse of the wave function. Before an observation is made there is nothing we can conclude from the wave function. This was the interpretation given by Neils Bohr and others for this strange experiment. So it is called the Copenhagen interpretation as Neils Bohr lived in Copenhagen. This observer dependant universe or reality was diametrically opposite to Einstein's objective reality, one that had nothing to do with any observer.

At this stage you may argue as Einstein did that well much the same thing happens when you toss a coin. You could get either a head or the tail and you don't know which, in advance. So what's so strange about it? But Einstein analyzed this problem in the spirit of classical physics. In effect what he said was that actually you don't know anything about the coin toss outcome because you do not know the values of all the parameters that are involved. There are so many hidden factors when you toss a coin. Firstly you don't know whether it is facing upwards or downwards at the instant of toss. Secondly you don't know the force with which the thumb strikes the coin. Thirdly you don't know how many tosses or spins it takes by the time it hits the ground. That again depends on the weight and shape of the coin and the atmospheric conditions. Now if you can estimate that you started with the head up and that it took six turns of the coin before it came to a rest on the ground then you can say that definitely you would get a head. If there were five spins of the coin before reaching the ground you would say definitely it would be a tail. So, in principle, you can predict the outcome. It is this missing information or knowledge that has created the probability.

Einstein went on to argue that similarly for Quantum systems there are a number of what he called hidden variables, about which we have no clue whatsoever. Because of these hidden variables there is the probability. He even designed an experiment to put forward this point. This experiment has become famous as the Einstein-Rosen-Podolsky paradox because it was put forward in a paper written with his colleagues Podolsky and Rosen and moreover it was paradoxical.

Now let's do this experiment in a simplified version . First we get into our Space Lab. Suppose we take two protons which as you know repel each other and put them together or nearly so. They are at rest to start with. Then suddenly we release both the protons. You can guess what happens. They move in opposite directions. One, proton1, let us call it, goes from point O to point A when we observe it. In the meanwhile the other has gone from point O to point B. Now let us say that our observation finds out the momentum of the first proton when it reaches A. This momentum we find to be in the same direction OA. Immediately we can conclude that the proton 2 at B at that very instant has the same momentum but in the opposite direction OB. We are quite sure about this because of the law of conservation of momentum. To start with the total momentum of proton 1 and proton 2 was zero as they were at rest. Therefore this would be the case for all time. When proton 1 is at A and proton 2 is at B the total momentum at that instant would still be zero. That's the point. So if one has momentum p (in direction OA) the other has momentum -p (in direction OB) which means it is in the opposite direction (Cf.Fig.3.9). Everyone will agree with this experiment, Einstein and Schrodinger included [21].

Then Einstein pointed out in that paper, look this experiment actually shows the shortcomings of Quantum mechanics because classically there is no problem with this experiment but Quantum mechanically there is indeed a major problem. And what is that problem? Let's suppose in a court of law Einstein and Schrodinger are debating this out. So Einstein says look

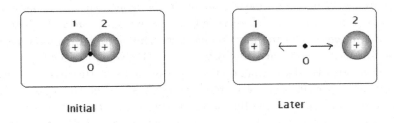

Fig. 3.9   EPR paradox experiment

you agree with this experiment but this experiment disagrees with your Quantum theory for the following reason. We are finding the momentum of the proton 1 when it is at A. You say that this can be done only by an observation. It does not have any pre existing value. Only when you strike it with a photon or whatever, will you conclude that its momentum is p in one direction. Good. What about proton B? How can you know its momentum without carrying out another experiment on the proton 2? Yet we have got its momentum without carrying out any observation which goes against what you are professing. That's what Einstein would argue. You are able to assign a value to an observable without carrying out any observation, and the observable is the momentum of proton 2 at the point B. Moreover, this information is obtained instantly, an idea that had disappeared with his theory of relativity, according to which there is a limit to the speeds in the universe – the speed of light. All this was therefore spooky, non-local as physicists call it, because it meant that information about the momentum of proton 1 obtained by an observation was transferred to proton 2 at that very instant.

Now Schrodinger had another take on this. His argument appeared a bit too ingenious, a bit too clever to many people. He said that there is no contradiction with Quantum mechanics! Why? Because when the protons 1 and 2 were at the same point O they formed one system and so were described by one wave function. After a certain time no doubt they have drifted apart but they are still described by a single wave function. Schrodinger called this the Non-separability Axiom. He further went on to say that this is one of *the* characteristics of Quantum Mechanics. So now how does this explain Einstein's question without contradicting Quantum theory? Schrodinger said that when you make a measurement of proton

1 you are really making a measurement on the combined wave function of proton 1 and proton 2. There is only one wave function and when you make the measurement it's just one measurement that gives you an answer to the momenta of proton 1 *and* proton 2. QED. No contradiction.

Neils Bohr also tried to explain this experiment [22]. His argument was: You see when you make a measurement on a system, firstly the system is a microscopic system. Secondly the measuring apparatus is a macroscopic apparatus. For example a microscope or a footrule or a clock or whatever. These are all large equipments. So really a measurement involves an interface of the micro world and the macro world. So its not really as if you can extract away the macro world which is described by classical physics, from the picture.

Einstein however argued that all this was too sophisticated. In his view, this was self-deception. The issue he said, is different. There are certain hidden variables which we have ignored because there is no way we can know them. The Quantum mechanical probability is caused because of this ignorance much like in the flip of a coin experiment. That was the crux of his argument. In other words, nature was not inherently probabilistic – there is indeed an objective reality. It is our partial information that is the culprit. So Quantum mechanics was incomplete.

Rosen, one of the authors of the paper later recounted how Podolsky the third author went to a New York newspaper after this paper was ready and disclosed that Quantum mechanics had been disproved. This was remember way back in 1935. When Einstein saw this report he was so upset and annoyed that he never spoke to Podolsky thereafter. In any case that was the great debate. In a sense there was only one jury, one way to resolve this debate between giants and that was very much in the spirit of physics. To perform an experiment to try this out and see who was right and who was wrong. The only catch was that technology till 1980 did not permit us to perform this experiment.

Thirty years after Einstein's 1935 paper a British theoretician associated with CERN Geneva deduced a few theorems. These are called Bell's inequalities after John Stuart Bell the CERN theorist. The inequalities in a sense could if true prove that Einstein's hidden variables do not exist. If not true then Einstein would be right. It was only in 1980 that technology had advanced sufficiently to allow the experiment to be performed because it required very fast electronic micro switches. That year a French experimenter Alain Aspect performed the experiment and – hold your breath!– to everyone's shock, at least all those who went with Einstein, it turned out that

Schrodinger was right! [23] Quantum Theory was right and Einstein was wrong. This meant that not just that Quantum Theory gave a few rules, a few algorithms for technology but also that as a theory, as a concept it was correct. The probability and uncertainty were built into nature. Like it or not, the universe is bizarre and defies commonsense! The uncertainty is not a result of our ignorance and our incapability or lack of technology for getting all the relevant information as Einstein believed. Even if we could get all the available information nature was inherently probabilistic. Einstein would say God does not play dice. But God, it appears, does play dice!

Curiously there is a story from ancient Indian folklore on this topic. The God Siva in heaven was consorting with his spouse Parvati and playing a game of dice. He threw the dice and there was a big sound. Parvati asked Siva what the sound was. Siva replied, well that's the sound of the birth of the universe. The next time round that Siva threw the dice, again there was a great sound and Parvati asked Siva what the sound was this time. Siva answered well that's the end of the universe. There are more dimensions to this story than appear on the surface [24]. In any case Quantum theory triumphed and we still had to explain all this uncertainty, the collapse of the wave function and questions like, "is the moon there when we do not see it", and so forth.

There was also another aspect to the EPR paradox that was very mysterious and continues to be a riddle. This has to do with the fact that a measurement at the point A yielded an observation and this information was instantly known at the point B as if information was travelling faster than the speed of light. In other words experiment appeared to contradict Einstein's theory of relativity according to which the information takes some time to reach B from A. Many arguments have been given to assure ourselves that the theory of relativity is not violated in this process. It is argued that while we do know what happens at B the instant we know what happens at A that very instant, nothing actually moves from A to B. We have got information which consists of two branches, one pertaining to the state at A and another pertaining to the state at B. The information has actually moved no signal, has moved no physical object. No particle or photon has moved from A to B so that there is no contradiction with special relativity.

This non local property happens to be a characteristic of nature however bizarre it may be. But it does not contradict relativity. Several arguments have been put forward in the past few decades and even today this argument is being bolstered with other arguments. It is an ongoing debate but finally the result is that this non local behavior of Quantum mechanics does not contradict special relativity. But notice that there is one assumption which has slipped in silently: we are able to speak of what happens at A and what happens at a distant point B because of what are called conservation laws, the law of conservation of momentum, the law of conservation of angular momentum and so on. That is the route. Now the laws of conservation have to do with what is called the homogenous nature of space and time. Space and time are everywhere and at all times the same and it is from this property that the laws of conservation emerge. They are used everywhere including in the EPR paradox. So this is a property , a non local feature that is assumed without further question.

As we will see in later chapters, this view of spacetime may not be the last word. Recent approaches in the studies of the unification of general relativity and Quantum mechanics use slightly different properties from this smooth homogenous spacetime. Therefore perhaps this underlying assumption of these holy laws of conservation are to be reexamined to reach a definitive answer that resolves the striking points that had kept Einstein, Schrodinger and a host of others arguing away all their lives.

The more practical aspect of all this has recently come to light. The EPR paradox is due to the correlation or what is called entanglement which is characteristic in Quantum mechanics which was the burden of Schrodinger's rebuttal of the EPR paradox. Entanglement speaks of two distinct objects forming in some sense a single quantum state. Not just two, there could be many more. It turns out that this property can have a wonderful role in building a new type of a computer, a futuristic computer called the Quantum computer [25]. Moreover this type of a supposedly instant transmission of information, is also a very very useful tool in these hypothetical Quantum computers. Much work has been done in this direction in the past three decades and there has been some progress. In any case we come to the idea of Quantum computers from a different angle. As computers are shrinking the chips are getting smaller and smaller. Is there a limit? The answer obviously is yes. If the chip comes down to the size of an atom or a molecule, then it no longer can be represented as a conventional computer. If it is small then we have to use the laws that apply at the atomic scale, that is Quantum mechanics. So Quantum computers are something of the future. Today we are slowly coming round to the view that in the not too distant future they may become a reality to make our conventional super computers look like nothing more than bullock carts!

There was another interesting development. In the 1950s a PhD student of the celebrated American physicist John Wheeler (who incidentally was also the supervisor of Richard Feynman) by the name Hugh Everett III in his doctoral thesis put forward perhaps the most crazy of all interpretations of Quantum theory and the universe itself. He said, look the wave function describes many possibilities. Now we are saying that an observation causes the wave function to collapse and pick out one of these many possibilities ignoring the rest, though in a very random way without any predetermined rule as to which of the possibilities it would pick up. Suppose this was not the case, suppose there was no collapse of the wave function but the wave function went on evolving realizing all the possibilities. Now how could that be? Everett's proposal was that the universe would evolve into a multiplicity of universes one for each of the possibilities. These are all "parallel" universes and it so happens that we are in one of them. In other words anything that can happen does happen though in its own universe. This was for many people a bit too much to swallow. And some were undoubtedly hoping that hidden variables would be discovered and there would be no need for these catatonic universes.

However in the seventies a new type of a thinking emerged. This was started by the work of H Dieter Zeh of the University of Heidelburg. He argued that the Schrodinger equation has a built in censorship which chooses one of the many possibilities. The work was carried out further by the Los Alamos physicist W.H. Zureck and a few others. Later these ideas became known as "Decoherence". Let us try to see in a simple way what all this is. The whole point is that all our ideas are in a sense what may be called two body ideas like the orbits of planets with Newtonian gravitation and Kepler's laws. Even if there are three bodies or objects as we saw and will see, things get far more complicated. So, much of the above arguments would be valid if let us say an electron was the only object in the universe or two protons were the only objects or they were completely isolated away from other objects. The moment there is a third object it is in some way interacting with the first two objects. It could be gravitational interaction, it could be electromagnetic interaction, whatever. The point is that there is interference and such an interaction is in itself an observation, though not necessarily originating in a conscious observer. Does it really matter whether the photon that makes the observation has been shot from a gun by an intelligent physicist or a chimpanzee or even an inanimate object like a radio active element? It doesn't. What really happens is, a photon goes and strikes the microscopic system or wave function. What does it matter how and from where the photon has come. So if we can ignore the environment then much of our theory would be quite okay as an argument. Unfortunately we cannot ignore the environment. There are so many objects all around even in a so called perfect vacuum. These objects could be particles, they could be electrons they could be neutrinos, just about anything. They are all the time interacting with each other and any quantum system. So such an interaction in whatever form would constitute an observation.

In other words the universe, every aspect of the universe is constantly under observation and that means continuously a collapse of any wave function is taking place. So even if Schrodinger's cat were neither alive nor dead that ideal situation would last no more than the minutest fraction of a wink. In no time some photon or something from somewhere would come and strike the wave function of the cat and the wave function would collapse into one of the two possibilities. This means that there would be a cat that is alive or a cat that is dead. The intermediate possibility is simply not possible in reality. This idea of decoherence has put at rest many of these paradoxes which riddled Quantum mechanics. At least it has been some sort of a palliative and brought down the temperature of the heat of the debates.

The entire saga of Quantum theory has some important morals to teach us. First we need to go with experiment because that's what physics is all about – about the real world. Second we need to have an open mind and consider all possibilities however absurd they may appear on first sight. Established ideas form blinkers which give scientists a one track view. It is very important to break out of this track and explore bold new ideas however contra to commonsense they may appear. That is another moral which Quantum theory teaches us. We will come back to these two morals later on because they are in a sense very important guiding principles.

Chapter 4

Time is running backwards isn't it?

"... there is the drying up of great oceans, the falling away of mountain peaks, the deviation of the fixed pole star ...."

– Maitri Upanishad.

4.1 Dealing With Billions

Twentieth century physics inherited the Newtonian concept of space and time. Some of these ideas remained while others were modified by Einstein's two theories of relativity. What happened was that time became less absolute and the concept of simultaneity got modified. We saw this earlier. A star explodes in Andromeda galaxy and there is a train crash on the earth one month later. For another suitable observer, the star explodes years after the train crash. Just as well, there could be a third observer for whom both these events took place simultaneously. There was now a new meaning for what had been recognized for ages, namely that cause should precede effect. That is, the father should be born before the son. That did not change.

According to relativity events which were not causally related, for example a train crash and a star in some distant galaxy exploding could take place interchangeably or even at the same time, depending on who the observer was (and his state of motion). But the son still has to be born after the father and not with or before him.

However, one of the old concepts which remained was that time was reversible. This means that in Newtonian mechanics the time t could be replaced by -t. This does not mean that time can go backwards, that we can grow younger by the day. The reversibility of time pertains to the underlying laws of physics. To understand this, suppose we enter our Space Lab and film, let us say two electrons approaching each other and then diverging after the encounter. The reversibility of the laws of physics means the following: Let us now run the film backwards. We would again see two approaching electrons which subsequently diverge. The point is, in this process do any of the laws of physics get violated? The answer is, no. The time reversed sequence of events would also be a perfectly legitimate scenario, something which can in principle happen in real life. There was a carry over of some of these concepts into Quantum theory as well. The universe is still reversible in time.

The funny thing is that all this goes against our experience. We don't grow younger as the days march on, nor do tall trees get shorter. In fact our everyday world is anything but reversible. The physicist would say that time has an arrow [26]. This contradicts the fact that the basic laws of physics are time reversible, that is do not have any arrow for time.

In our real world we can put this in a different way, in terms of what is called entropy. This is a much used and even fashionable word. It features prominently in what is called the Second Law of Thermodynamics, which states that the entropy keeps increasing. (The first law is a restatement of the well known Law of Conservation of Energy). Entropy, roughly speaking, is a measure of the disorder in a system, and every housewife knows the second law, at least in principle. She tidies up the house with everything in its right place. But as the day goes on many things end up in the wrong place. These things could be the children's building blocks or coffee cups or clothes, just about anything. In other words the disorder in the house increases. She has to put in a lot of effort at the end of the day to pull everything back to its right place and restore order. Without this effort however, the disorder would keep increasing.

We must bear in mind that in the nineteenth century, a new discipline was born, which also had a new ethos——rather than being an abstract study of the universe, this new discipline, Thermodynamics was a child of the industrial era triggered off by, amongst other things, the invention of the steam engine. Let us spend a little time here, as we will encounter these ideas later. In the words of Toffler [27]:

"In the world model contributed by Newton and his followers, time was an after thought. A moment whether in the present, past, or future, was assumed to be exactly like any other moment...

"In the nineteenth century, however, as the main focus of physics shifted from dynamics to thermodynamics and the second law of thermodynamics was proclaimed, time suddenly became a central concern. For, according to the second law, there is an inescapable loss of energy in the universe. And, if the world machine is really running down and approaching the heat death (the final stage when the universe runs out of energy), then it follows that one moment is no longer exactly like the last. You cannot run the universe backward to make up for entropy. Events over the long term cannot replay themselves. And this means that there is a directionality or, as Eddington later called it, an "arrow" in time. The whole universe is, in fact, aging. And, in turn, if this is true, time is a one-way street. It is no longer reversible, but irreversible.

"In short, with the rise of thermodynamics, science split down the middle with respect to time. Worse yet, even those who saw time as irreversible soon also split into two camps. After all, as energy leaked out of the system, its ability to sustain organized structures weakened, and these, in turn, broke down into less organized, hence more random elements. But it is precisely organization that gives any system internal diversity. Hence, as entropy drained the system of energy, it also reduced the differences in it. Thus the second Law pointed toward an increasingly homogeneous——and, from the human point of view, pessimistic——future.

"... time makes its appearance with randomness: 'Only when a system behaves in a sufficiently random way may the difference between past and future, and therefore irreversibility, enter its description.' In classical or mechanistic science, events begin with 'initial conditions,' and their atoms or particles follow 'world lines' or trajectories. These can be traced either backward into the past or forward into the future. This is just the opposite of certain chemical reactions, for example, in which two liquids poured into the same pot diffuse until the mixture is uniform or homogeneous. These liquids do not de-diffuse themselves. At each moment of time the mixture is different, the entire process is 'time-oriented'."

A steam engine, for example, loses heat (and steam) into the atmosphere and even to the body of the engine, which then radiates it to the atmosphere. That is, the heat gets "dissipated" into the billions of molecules in the atmosphere. The entropy increases. This increase gives direction to time: The entropy increases with time. If all this "lost" heat could be reclaimed, then the efficiency of the engine could be boosted–the same amount of coal could take the engine much farther.

In a sense these "thermodynamic" ideas were anticipated in the nineteenth century even in the field of mechanics, through the work of Poincaré and others, working in the field of celestial mechanics rather than industrial machines. Were the orbits of the planets or other celestial objects really unchanging in time? Newton would have said, "yes" but Poincaré realized that celestial mechanics had been worked out under the banner of what we called the two body problem. The orbit of the earth round the sun, for example, would be more or less unchanging, if the Earth and the sun were the only two objects in the universe. Even with a third planet, we have to consider the three body problem, which as Poincaré realized had no exact solution [28]. In fact Poincaré submitted an essay on this topic and was awarded a gold medal in a competition in honour of King Oscar II of Sweden in 1887. This essay contained a serious error, as he later realized himself, while correcting the pre publication proofs of the article! He had laid the ground for what has subsequently come to be known as chaos theory. As Prigogine was to say much later [29]:

"Our physical world is no longer symbolized by the stable and periodic planetary motions that are at the heart of classical mechanics. It is a world of instabilities and fluctuations..." We will return to this later. The whole point is that in a very orderly universe, time would be reversible and there would not be any increase in disorder or entropy. Strictly speaking this would not be possible if there were more than two objects in the universe, which then becomes dissipative. We lose energy on a one way street.

Let us now perform another experiment in our Space Lab to illustrate all this even more clearly. We take a large container, let us say a cube, which is air tight (Cf.Fig.4.1). There is a partition right in the middle of the cube. However there is a tiny hole, or even a few holes in the partition. Let us now fill up half of the compartment, let us call it part A with molecules of air. Typically there would be trillions of trillions of molecules in A. In B however there are no molecules - (Cf.Fig.4.1(a)) they have all been pumped out. Now let us see what happens as time "passes". Many of the molecules in A will hit the partition and get reflected back. By chance however one or two molecules might pass through the hole, that is dissipate in to the compartment B. As time passes, this keeps happening, except that one or two molecules which have entered B would also pass through the hole in the partition and return to A. This happens much less frequently because there are much fewer particles in B compared to A, which started out full if you remember.

Let us make all this very specific. Suppose, to start with there are a thousand molecules in A and further each of these molecules have been numbered. By observing you can know that by chance molecule number

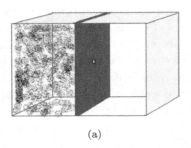 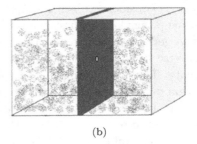

(a) (b)

Fig. 4.1   Diffusion of molecules

23 has escaped from A into B, then molecule number 87 for example. And so on. All this happens in a very random manner. So equally well, what is happening to our compartment can be described in the following way. There is a person sitting with a directory (or counter) of random numbers between one and thousand (the total number of molecules in A). The person picks out a random number 27 for example. He then pushes out the molecule number 27 into B. Next he picks out the random number 95 and does the same thing. A second person would also be doing the same thing with the molecules in compartment B – but his disadvantage is that there are very few molecules in B to start with. So most of the random numbers he picks up would not have the corresponding molecule in B. This would give an equivalent description of what was happening. In due course you would find that the number of molecules in A and B level off - there would be about five hundred in A and five hundred in B (Cf.Fig.4.1(b)). Sure a few molecules from B would be drifting over into A and vice versa, but on the whole there would be approximately an equal number of molecules in either compartment. Both the persons with the random numbers would now find equal number of molecules matching their random numbers. More or less.

We have here a scenario where time has exhibited an arrow, a bias. To start with there were a thousand molecules in compartment A. Though molecules from A were leaking into B and vice versa, on the whole there was a predominance of molecules going into B so that compartment A got depleted at the expense of compartment B. This process clearly is not reversible. You have to bear in mind that the laws that we have been using are all reversible laws, as far as the motions of the individual molecules are concerned. But the nett result is not symmetrical or reversible [29].

Let us consider two other simpler real life situations which bring out the same principle. The first one is where a plate of China falls from the table and gets smashed into a large number of pieces. We don't encounter the reverse situation where a large number of pieces come together and form a plate. The other example is, if you take a drop of milk or ink or a cube of sugar or whatever and put it into a tumbler of water, what happens is that the molecules of water which are constantly jiggling, buffet the molecules of the drop of ink and spread them out. The drop disappears. Sooner or later the tumbler of water gets almost uniformly colored with ink. The concentrated molecules in the drop have diffused nearly uniformly into the surrounding liquid. As there are no forces or any special mechanism acting on the drop, can we expect that the molecules of ink in the uniformly colored liquid would, due to the same jiggling of the water molecules combine back and give us the original drop of ink? Or that the sugar unmixes back into the cube? That never happens.

Both these are examples of the increase in entropy. We started out with a highly ordered situation, where the entropy was very low and ended up with very high entropy or disorder. The physicist would say that it is highly improbable that entropy would decrease and we recover the drop of ink from the blue liquid, whereas the increase of entropy is by far the more probable of the two. We encounter what is more probable. We do not, in the normal course, meet a person with one blue eye and one brown eye, because that is such an improbable event.

So the picture that emerges is that from underlying symmetrical laws we can get a statistical non symmetrical outcome. There is no contradiction with the fact that time could go either way in the laws of physics. There is no arrow of time to start with. A molecule can equally well go from A to B as from B to A, but when we consider billions of molecules and the fact that we started out with many many more molecules in A compared to B, then we end up with a non symmetrical process where molecules essentially have leaked one way from A to B. This is the arrow of time. The same explanation holds good for why a drop of ink permeates into the surrounding water and the reverse doesn't happen. It is most unlikely.

Similarly, there is an electrodynamic arrow of time. This is a situation with apparent time asymmetry where the electromagnetic wave reaches outwards from a given electric charge. For example suppose there is an electron at a point A. We now introduce another electron at a point B. Physics tells us that an electromagnetic wave leaves A, moves with the speed of light and reaches B. If we jiggle the electron at A, in response, a little later, the electron at B also starts vibrating. The effect comes after the cause. There is an arrow of time. But even here we can invoke the statistical picture: There is a formulation of electrodynamics put forward by Wheeler and his celebrated student Richard Feynman around 1945, in which there is a time symmetry. It is not that a wave goes from an electron outwards. Rather waves go from the electron outwards towards other charges and waves come in equally from the other charges to the electron. It is a bit like molecules diffusing from compartment A to compartment B and vice versa: Statistically, over all there is this arrow of time, with the final electromagnetic waves emanating from the charge [30].

This was also Einstein's belief. He swore by time symmetry, on the one hand. But realized that electron B above will jiggle only after we jiggled the electron at A, as we saw. Otherwise his theory of relativity would be contradicted and effect would precede the cause. He took his inspiration from what happens in heat. As you know heat flows from a body with higher temperature to a body with lower temperature not the other way round. Einstein realized that there is a statistical explanation for this too, exactly of the type that we have seen for compartment A and compartment B.

Unfortunately these ideas which were put forward by the great Austrian physicist Ludwig Boltzmann in the nineteenth century did not evoke an encouraging response from peers. Boltzmann was the first to put forward the law of entropy, that there is an arrow of time, the one way street increase in entropy or disorder. All this stemming from time reversible Newtonian mechanics. Boltzmann's critics pointed out that, look what you have got here is an arrow of time from a no arrow or symmetrical mechanics. Boltzmann was so hurt by all this sustained criticism that finally he committed suicide. Ironically, shortly after his death, scientists began to appreciate these ideas.

In any case, we must bear in mind that in the above description, we have gone beyond Newtonian mechanics. We have introduced a new element – ideas of statistical mechanics, dealing with a large number of particles and concepts like probability. In Newtonian mechanics, it was possible in principle to predict exactly the position and speed of a particle at a later time, if we knew where it was in the past, how fast it was moving and the forces on it. When we deal with billions of particles, this technique is impractical. We deal with averages, probabilities and so forth.

Today we do not find anything so wrong in all this. Time is symmetric at the micro level. The asymmetry or arrow comes from the asymmetric initial conditions. For example, to start with, compartment A contained a thousand molecules while compartment B did not contain any. If we reversed B and A to start with, then the opposite would happen. On the other hand, when A and B have the same number of molecules, then as we saw, the semblance of symmetry returns. The same number of molecules leak into B as into A.

If we look back, we started with equations in Newtonian mechanics or Quantum mechanics, where you can replace time going one way by time going the other way and still get perfectly correct and meaningful equations. That is time is symmetric. That was our starting point. Interestingly however, in the micro world there is one exception, a remarkable exception to all this. Only one experiment contradicts this time symmetry and that is the decay of an elementary particle called the K^0 meson which takes place in a non symmetrical fashion. The K^0 meson or kaon can be considered to be oscillating between two states, A and B let us say. These are like two different particles, with different properties. A decays into two particles called pions, and this decay takes place very fast. B on the other hand lives much longer and decays into three of these pions. This much longer life time is still very small, less than one billionth of a second. So after a sufficiently long time, there should be hardly any A particles left in a beam - mostly B particles would be left and these would be decaying into three pions each. However, such two pion decays were observed as early as in 1964. This is an asymmetrical scenario and indirectly constitutes a violation of time symmetry. It's a very small exception but who knows there may be something very profound here. In fact the great Oxford mathematician and physicist Roger Penrose had this to say: "the tiny fact of an almost completely hidden time-asymmetry seems genuinely to be present in the K^0-decay. It is hard to believe that nature is not, so to speak, trying to tell something through the results of this delicate and beautiful experiment" [31].

Apart from this exception time at the micro level is symmetric. It is only when we come to the macro level that time becomes asymmetric in the statistical sense. So time symmetry without an arrow in the small can and does lead in the large to time asymmetry, that is with an arrow. As we are all macro creatures, that explains why we don't grow younger by the day, or why trees don't get shorter as the years roll on. On the large scale, there is an asymmetry, an arrow of time.

Or is this the whole story? Have we done our experiment with the two partitioned chambers of the molecules all the way through? Or have we abandoned the experiment as soon as both chamber A and chamber B came to have the same number of molecules? Have we left the experiment at that stage concluding that time goes one way?

Suppose we go back to the Space Lab and that experiment and keep on observing what happens. Well may be for some time a molecule jumps in from A to B and equally from B to A. So on the whole, for some time the chambers A and B might continue to have the same number of molecules. But after some time there would be a fluctuation. Remember fluctuations are random and we did the experiment with a book of random numbers. In this randomness anything can happen however improbable. Otherwise it is not random, if it becomes a rule, a law that only something can happen and not anything else. For example, let us go back to the two random number directories. It may so happen that for a very large number of times, we may pick random numbers that do not correspond to any molecules in chamber A, compared to those that do. Then, the nett influx of molecules into A far exceeds the exodus from A. So because of the random nature in which molecules are swapped it may so happen that at some point of time the chamber A may again have more molecules compared to chamber B. In fact so many more molecules that we may be almost back to where we started: with a full chamber A and an almost empty chamber B. This would happen sooner or later.

There is a famous theorem proved by the great Poincaré more than a hundred years ago. What Poincaré proved was that if you start with a finite amount of space with no energy being lost or gained in that finite amount of space, and if there were a certain configuration of molecules at some point of time then, after a sufficiently long interval of time you would get back that same configuration or something practically the same. This applies also to our two chambers A and B put together. You would get to a certain point of time if there was no exchange of energy between the ambient and chambers A and B put together when you would be back with the starting situation where there are overwhelmingly more molecules in A compared to B. In fact may be there would be practically none in B.

Now we can say that look, where has the arrow of time gone? We started with molecules in A and just about none in B and after a sufficiently long amount of time we have come back to the original status. This means that time is symmetric. It is not that it goes only one way. It is almost as if, provided we wait long enough, the molecules of the ink drop which permeated into the tumbler of water, all recollect back into the drop. So on the whole one would be tempted to say time is again symmetrical or reversible even from this statistical point of view. That is, we might as well say that time is just as well, running backwards!

There is of course one important caveat: We must ensure that none of the other conditions change in the meanwhile. That is the catch. Unfortunately that may not be the case. For instance if one would have to wait for some ten billion years for the molecules of the ink to collect back into a drop, then who knows, in the meantime the universe itself in its present form might come to an end, and no longer we would have the same laws operating. That is entirely possible.

There is another aspect. We are talking about the probability of molecules to diffuse or undiffuse. That is not the only process in the universe. There are so many other processes with different probabilities which keep changing with time. For instance the radioactive decay of elements. Or may be the molecules meanwhile chemically combine to form new molecules. So it is not clear. May be time on the whole is symmetric but as we are saying, providing other conditions don't change. This may not be probable, and time after all may end up having this arrow. It may be not so symmetric. Wheeler on the other hand was confident, "Past and future are symmetrically related" [32]. (This sentiment was shared by Einstein). Now Wheeler said this in the mid nineties when the picture of the universe was that of a birth in a big bang. Then the universe would evolve but again it would collapse and end up in a big crunch. All the matter would implode inwards and there would be a big squeeze, the opposite of the big bang. A few years after Wheeler wrote this, the scene dramatically changed, and it is no longer very clear as we will see later, that what Wheeler said holds anymore.

4.2   Time, the Eternal Enigma

From a classical, that is non Quantum point of view, we might be tempted to think in terms of all space or the entire universe at one instant of time. As this instant progresses, the universe evolves. To put it more dramatically, let us think of the entire universe as resembling an apple. At a given instant of time what we perceive of the whole universe is a thin slice of an apple, the position of the slice representing the particular instant of time. At another instant, for example a little later we would have to deal with a thin slice just above the earlier slice (Cf.Fig.4.2). This is how Newton might have described the progression of time. But then Einstein came and mixed up space and time in his theory of relativity.

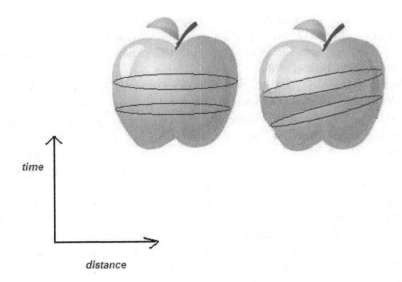

Fig. 4.2   The universe as an apple

In this relativistic picture, we have to deal with events and intervals. For example we might describe all the events in the universe which took place at let us say 6.00 p.m. by our clocks. Of course there are intervals between these various events. In our previous example of the apple, these intervals between the various events would be the distances between them, or the distances of various points on a given slice from the centre. Now however, the points on the slice of the apple are the events and the intervals are spacetime intervals: they contain both space and time. It is as if the slice gets tilted. The distances between the points on the slice are no longer absolute, in this new description. These distances would depend on the observer. From a series of slices a certain sequence of slices is then extracted out – remember each slice has not just a spatial extension, but rather a spacetime extension in relativity. By abstracting out a sequence of slices we really get a description of how the universe is evolving - this sequence also contains a description of the laws of physics. But there are rules. For example if we removed certain slices from the sequence and inserted instead other slices, this could mean that the universe evolves without obeying the laws of Newton. Or it could mean that the father was born after the son because, in the mix up, we put the son slice before the father slice. In particular a particle with no force acting on it could in a special case oscillate. This would go against the known laws of physics. All this would still be forbidden.

Unfortunately even this purely classical description is inadequate. A major problem arises when we encounter what is called a spacetime singularity in general relativity. Such singularities occur inside a black hole for example, or at the very instant of the big bang which led to the creation of the universe. Black holes have captured everybody's imagination and are a consequence of general relativity. To understand what these mindboggling objects are, let us perform a simple experiment.

We throw a stone up. It goes to a height and falls back to the earth. Next let us throw it up even faster. This time it goes to a greater height and drops back a greater distance away. The moral is, the faster or more forcefully you throw it up, the greater the height the stone attains and the farther it falls. Now if you could throw up the stone with a speed of around eleven kilometers per second or forty thousand kilometers per hour, which is an enormous speed, what will happen is that the stone will go so high, it would have crossed much of the earth's atmosphere. At that height it begins to fall back but at a distance that is so far away, that it is just beyond the earth! (Cf.Fig.4.3). In other words the stone would be orbiting the earth as a satellite. The speed of eleven kilometers per second is called the escape velocity for the earth. All satellites have to be launched by rockets which attain this speed in order that they fall beyond the earth, or to put it another way keep on falling round the earth in their orbit.

This escape velocity is the correct value for the earth, but if you were on the moon or Mars or elsewhere the escape velocity would be different. This is because the escape velocity depends on two separate factors – the mass and size of the planet. For example if suddenly the earth contracted to a fourth of its diameter, the escape velocity would double and the rocket would have to be given this minimum punch in order to put a satellite or whatever into an orbit.

If on the other hand, the earth collapsed to less than the size of a mustard seed, then it is possible that the escape velocity can be three hundred thousand kilometers per second, the speed of light. In this case you would have to blast off a rocket to attain this speed to put anything into this shrunk earth orbit. But how is that possible? We all know that the speed of light is the maximum. So what happens in this case? Nothing, not even light can escape from such an object. That's why it's called a black hole, because

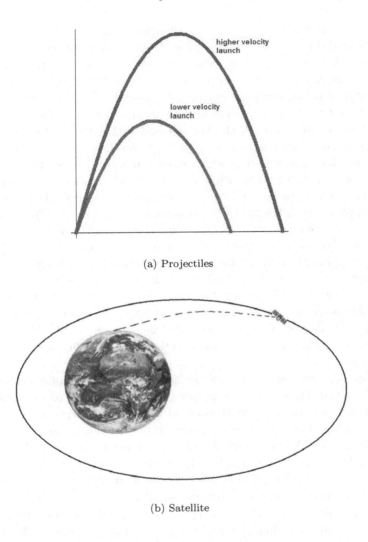

(a) Projectiles

(b) Satellite

Fig. 4.3   Projectiles and Satellite

we can never see it. This name was coined by Wheeler. Now let us switch our perspective and look at a black hole from outside.

Suppose our Space Lab is crashing into a black hole. As it approaches closer its speed increases. The Space Lab spirals into the black hole with ever increasing speed. At a certain distance from the centre of the black hole, called the horizon, the Space Lab attains a velocity very close to that of light. It is not possible to conceive what happens once this horizon is crossed. Naively we would expect that the velocity would be greater than that of light within the horizon as seen from outside but that is meaningless. General relativity tells us that spacetime gets stretched and curved more and more as we approach the horizon and within the horizon spacetime is so distorted that our astronaut in the Space Lab would not be able to communicate with us any more. You can think of it in a simple manner. A gun man is firing away, while his van is rushing away in the opposite direction. Once the speed of the van attains the speed of the bullets, will these bullets reach their targets? No chance! Similarly no communication from our astronaut would reach us – he is rushing away at the same speed as that of the electromagnetic radiation he is beaming out.

More importantly right at the centre of the black hole, there would be a complete breakdown of the laws of physics – in a sense spacetime, would loose its meaning. This central point is called a singularity. In these special situations, the laws of physics are meaningless and there is no recipe for dealing with or accounting for such singularities. It is a bit like being exactly at the North Pole. Any direction could be the East or West or whatever.

Further problems arise if we consider the effects of Quantum theory. Without Quantum theory we at least were not dealing with probabilities and uncertainties. Now we have to contend with such concepts as this or that may probably happen, something alien to pre Quantum thinking. Stephen Hawking, the celebrated British physicist realized that rules of Quantum mechanics have to be used in the theory of black holes, even though black holes are a beast from the stables of general relativity. He showed that though no information or radiation should reach us from black holes according to general relativity, this is not true strictly speaking. Black holes do radiate but according to Quantum ideas. That is because a photon may still have a small probability of leaving the clasp of the black hole. They are not quite that black.

In Quantum theory, as we saw, when we approach smaller and smaller space or time intervals or spacetime intervals the momentum and energy keep shooting up. This if you remember was the content of the Heisenberg Uncertainty Principle. Quantum mechanics gives us another bizarre result. There is an interval that is so small – typically it is less than one thousand billionth of a centimeter or one trillion, trillionth of a second inside which our concepts of space and time break down. This interval is called the Compton interval – we call it the Compton wavelength or the Compton time. Inside it, again our concepts of physics break down, including concepts of causality. It is almost as if speeds greater than that of light are allowed within these intervals. As Wheeler puts it, "no prediction of spacetime, therefore no meaning for spacetime is the verdict of the Quantum Principle. That object which is central to all of Classical General Relativity, the four dimensional spacetime geometry, simply does not exist, except in a classical approximation."[33] Think of the apple of (Cf.Fig.4.2), but this time it is shrunk to less than a trillionth of a centimeter in diameter. Then the "father" slice and the "son" slice in the apple can indeed get mixed up! We will return to this scenario later.

Ultimately, as we can all understand, time is associated with change. In a changeless universe, there would be no time. A shortcoming of many of our ideas is that these theories of time already imply a notion of time, or more generally, change. The question that crops up is, what type of a change do we consider? The time which we usually encounter is based on small, incremental or continuous change. A child gradually becomes old. This does not happen overnight. What is left unsaid is that the shorter the interval of time, the smaller is the change. Try measuring the height increase of a boy every one second.

However, as we just now saw, once we consider the microscopic world and so allow in Quantum theory, all these considerations come under the scanner. This is again at the minutest distances possible. The old ideas of cause and effect and continuous change breakdown. Within these microscopic intervals, in a sense we have to deal with a sequence of events, which are no longer connected by the laws of physics. You could say that these events are chaotic. Worse. The very concept of time in our usual sense breaks down: The celebrated Nobel Prize winning physicist, Eugene Wigner pointed out that time is measured with macroscopic devices – some form of a clock that is made of several molecules. In his words, "... the inherent limitations on the accuracy of a clock of given weight and size, which should run for a period of a certain length, are quite severe. In fact, the result in summary is that a clock is an essentially non-microscopic object. In particular, what we vaguely call an atomic clock, a single atom which ticks off its periods, is surely an idealization which is in conflict with fundamental concepts of measurability."

Now let us consider extremely small intervals of subatomic dimensions, the Compton scale for example. Our clock, however small, is still much too large to "fit" into these small intervals and give us any meaningful information. It is only when we consider an aggregate of such microscopic events that the harmony of physics or the universe is restored. It is rather like a huge crowd chanting a slogan. Different people may chant it at slightly different times with slightly different pitches and so on. We cannot really analyze this with precision. But these individual differences are smoothened out when we hear the chant of the crowd as a whole. In other words to borrow a phrase from the celebrated Russian-Belgium physicist Ilya Prigogine, "Order emerges from Chaos" [34].

My own approach has been that time at a micro level could and does run backwards and forwards. This can be modeled by the example of what is called a Brownian or Random Walk (Cf.Fig.4.4) [35]. In this case a person walks along a straight line, but sometimes he takes a step forward and sometimes he takes a step backwards. Each step is of unit length, let us say. However all this happens totally at random. Think of it like this. He has a book of random numbers. He picks out a number at random. If it is even, he takes a step forward. If it is odd, he takes a step backward.

Fig. 4.4   A random walk

Where will the person be after several steps? We might be tempted to think that the person would be almost where he started. That would go against randomness as we argued! It would mean that the number of forward steps equal the number of backward steps. That would be the case if he flipped a coin and walked forwards when "heads" turned up and backwards for "tails". The flip of a coin has a definite probability associated with it – on the whole we are likely to get as many "heads" as "tails". Not so for random steps. Actually the person will find himself some distance away, though this may be forward or backward. At this stage the words forward and backward loose meaning. Forward and backward with respect to what? All that matters is that the person has moved away from the starting point though this distance would not be the total number of steps he has taken (multiplied of course by the length of each step). It is this distance which we characterize as the passage of a certain amount of time if each step denoted, let us say one minute or whatever. Remember that each step is actually a progression in time. In this picture time is not a definite change that can be attributed to each and every constituent of the universe. Rather it is like a wholesale concept, like the chant of the crowd. Or to give another example, it is like the average speed of molecules in a huge collection of molecules. Time now takes place not in a continuous manner, but rather in discrete or jerky jumps.

I would like to emphasize though that there are two aspects in this approach, as we just now saw. Let us go back to the example of molecules colliding at random in a container. There is what is called a mean free path, an average distance that any molecule covers inbetween collisions. Remember that there are billions of molecules and billions of collisions. So this average distance is a rough estimate indeed. This average distance is like our unit step in the random walk. There is also the distance covered in a number of such steps, some backward, some forward. Applied to time, this would translate into the unit of time and the total time elapsed. We will return to this theme.

Let us conclude this line of thought for now. We can see how information comes in, directly or indirectly. Even in relativity, time depends on the observer - his velocity - though in a deterministic way. In Quantum theory the observer has a more critical role – he interferes with and influences observations. In the statistical approach, probability comes in, but in a slightly different context, and this probability is closely linked to time. If there were no probability, if the events were all certain to take place or not to take place, in a sense there would be no time.

Take the plate breaking example. If in advance, the exact shapes of the would be pieces were prescribed, the chances for this happening would be practically zero – such an event would almost certainly not take place, as too much information is prescribed. Similarly, in the two compartments example, if the hole in the partition is fitted with a one way valve, one that allows particles to only go from B to A and not the other way, then A would remain as it was and no molecules would end up in B as if time did not pass. The point is, in these artificial examples, we have "distorted" the passage of time - the probability must be spontaneous. As every insurer knows, there must be the element of genuine chance for insurance to be meaningful! There cannot be too much information in advance, to break the spontaneous flow of time.

Today's ideas are that at a time in the future, the reverse of everything in time may never take place. This is because, the universe may not slow down to a halt and collapse, to start the cycle all over again. But scientists do not give up so easily. A view to which many are converging is that our universe is just one amongst many such universes – may be trillions of trillions of them, or more. And surely, in some other universe, the reverse phenomenon of time would take place. So if we stretch our canvass wide enough, we have situations of time running forward and also of time running backward!

Another point which has troubled many physicists, is the role of consciousness in the concept of time. Undoubtedly the two are linked though the nature of this link is as yet extremely unclear. But let us look on the universe as a collection of independent events. These events represent constant fluctuations that are taking place like the steps in a random walk. Fluctuations are random changes. These changes are taking place within us, the observers too. The brain consists of so many billions of restless molecules called neurons which are in incessant movement. Latest neural brain research points to the fact that these billions of motions in the brain which give rise to consciousness, are themselves chaotic as Antonio Damasio, a frontline brain researcher points out. So, perhaps, if the universe had been deterministic, then these neurons would not have this chaotic motion, and – who knows? – there may not have been consciousness itself! In any case they are outside the control of the consciousness which they create [36]. From this point of view consciousness itself is a part of the environment, a part of the ceaseless change, which it tries to comprehend.

Chapter 5

On a collision course

"Bodies are formed of ultimate atoms in perpetual vibration"

– Kanada
Ancient Indian thinker C. seventh century B.C.

5.1 Bits of Atoms

The Large Hadron Collidor installed in an underground tunnel of twelve kilometers perimeter or circumference in Geneva is the most expensive scientific experiment on the earth todate. It costs some four billion dollars. It was finally launched in Summer 2008 already behind schedule. To the jubilation of everyone, particles started flying about. But even before the clapping and the sound of champagne corks could die down, there was an unexpected shock. A short circuit ripped through part of the facility and disabled it. Well, repairs are underway. The bill is likely to run into twenty five million dollars. It now looks like meaningful physics data can come out only in 2010 at the very earliest.

For many years scientists had been keenly awaiting the Large Hadron Collidor or LHC, like it was a saviour. This is because particle physics stagnated after the early seventies, at least as far as theoretical and conceptual breakthroughs go. It was recognized that we needed more experiments to unlock more secrets and break out of the impasse. For example in the sixties a very important prediction was made. This had to do with a particle called the Higgs boson. It was postulated by a British physicist, Peter Higgs. The Higgs boson has also been called the God particle. This is because it is believed according to the accepted theory that the Higgs boson gives mass to all the particles and mass as you know is one of the most fundamental properties.

However though we have lived with this idea for some forty five years, the Higgs boson has not been fished out to-date. A few years back, a prominent British daily's headline screamed, "Is the God particle dead?" This is one example.

Quantum theory has predicted several such bizarre, credulity testing particles. The magnetic "monopole" is another example. We all know that there are positive charges, as with protons, or negative charges, as with electrons. Moreover, these can lead separate lives. With magnetism, there are similarly, the North magnetic poles and the South magnetic poles. But that's where the similarity ends. Nobody has ever found a way of isolating North magnetic poles or their Southern counterparts. They are always together in inseparable pairs. Physicists took it for granted that they cannot ever be wrenched apart. Not so, said Dirac nearly eighty years back. He argued that Quantum mechanically we could have these "monopoles", though under very special circumstances.

Then, another Quantum Mechanical theory proposes that there are what are called "supersymmetrical" particles – as many of them as there are ordinary particles. This default theory answers some awkward questions. But then, where are these new particles?

There were a number of law suits which tried to abort the launch of the LHC because there were fears in some uninformed quarters that the LHC would recreate something like the Big Bang which triggered off the universe's titanic blowout. Could this recreation trigger a major catastrophe that could rip apart the earth? Some European scholars, without a background in physics, asked me fearfully, if some newspaper reports about the impending catastrophe were indeed credible! In any case we have to wait at least till 2010 to find out! May be further beyond, for new discoveries from the LHC, discoveries which are so badly needed to infuse fresh life into physics.

What are these elementary particles as we call them for which so much money has been splurged on building giant machines or accelerators as we call them or collidors. This topic has also bagged so many Nobel prizes over the years. Surely it must be something very important. Well, the story goes back to several centuries before Christ, to India and somewhat later to the Greeks. There was the belief which seemed to go against everything that we observed that all matter is made up of some ultimate sub-constituents which the Indians called "anu" and the Greeks called atoms.

The idea that material which we see around is made up of much smaller invisible subconsituents, goes against our commons sense. For example as we saw Kanada the great Indian thinker who lived perhaps in the 7th or 8th century BC had put forward this idea [37]. It is also to be found in ancient Indian literature . For example the Vishnu Purana proclaims "How can reality be predicated of that which is subject to change, and reassumes no more its original character? Earth is fabricated into a jar; the jar is divided into two halves; the halves are broken to pieces; the pieces become dust; the dust becomes atoms." This was of course a philosophical discourse asking the all important question, "So what is reality?".

In the Greek world similar ideas came into play a little later. For example Lucippus seems to have started the tradition of atoms being the ultimate subconstituents of matter. And this carried on over the centuries with different scholars. For example Democritus too added his own voice. It was only two thousand years later in the second half of the eighteenth century that the old ideas made a come back, but this time backed by scientific arguments rather than mere brilliant conjectural insights. A number of what we may call scientists today, were involved in the resurrection of the idea of atoms.

There was Lavosier in France and Richter in Germany. In England there was John Dalton. All these studied the way in which different chemicals combine in reactions and took a hard look at the proportions in which they went in. Based on these studies they put forward a list of chemical elements. Much later Mendelev in Russia systematized this knowledge in his famous periodic table of elements. When Mendelev first formulated his table, there were just about six known elements, compared to over a hundred today! The important thing is that elements always combine in the same fixed proportion to give other compounds.

This suggested the brilliant atomic hypothesis: First, every piece of material is a collection of simpler parts called molecules. Molecules could not be seen of course, they were much too small for that. But the physical and chemical qualities of each particular molecule gave the character to the combination. And molecules were made up of atoms – sometimes identical atoms. Well the atomic hypothesis was just a sort of a guess. The fact was that the subunits making up the molecules are the basic elementary building blocks of chemicals. These were the atoms. So, not molecules but atoms were the ultimate building blocks of all matter. On this basis a number of successful predictions were made by the end of the nineteenth century.

Prout, a Britisher was one of the earliest pioneers of this theory. Let's take a look at his theory. Early in the nineteenth century Prout had suggested that the hydrogen atom is the basic unit for all elements and that other atoms were multiples of the hydrogen atom. This was a very bold idea, definitely one of the brilliant insights of that century. One could now think of different elements being formed with the building blocks of the hydrogen atom. These new elements would have different properties compared to the hydrogen atom of course. Almost a hundred years after Prout, measurements were made to test his hypothesis. He was nearly correct, but not quite [38].

In the nineteenth century, for much of the time there were attempts to break down the subconstituents of matter. These attempts were based on supplying more and more heat to the material so that the heat would tear them apart. This technique could go only so far.

A new technology provided the next breakthrough. This came later in the nineteenth century – electricity. Now electrical discharges were triggered off in low pressure gases and this helped the atoms or molecules to rip apart. Using this technique the British scientist J.J. Thompson in the 1890s discovered the electron, a charged particle. The name electron itself comes from the Greek word for amber. It was known from the beginning of history that if amber were rubbed it produced some sort of a peculiar effect which today we recognize as static electricity. Thompson could identify the electron. But the gas or matter itself did not have charge. So it was concluded that there were the negatively charged electrons on the one hand constituting matter and there were also positively charged subconstituents which neutralized the electron's negative charge.

As we have already seen the experiments of Rutherford followed early in the twentieth century and Bohr put forward his model harnessing the Quantum hypothesis of Max Planck. But returning to the atom, we had the electron as a constituent and then there was the atomic nucleus itself. The picture which appeared attractive was that the nucleus was made up of positively charged protons. But this turned out to be incompatible with several experiments. That is because, if we measured the mass of atoms with the number of electrons and protons, that didn't match.

In 1932 the British experimentalist James Chadwick discovered another particle, a neutral particle, the neutron which was not equal to a proton plus an electron. It was something else. So atoms had electrons orbiting like planets and then there was a central nucleus like the sun. This nucleus consisted of protons and neutrons. That seemed to explain just about everything.

However other particles or elementary particles burst forth into the open from different routes. For example as we saw, in 1928 Dirac had put forward a theory of Quantum mechanics which included special relativity in addition. And the equation was found to represent an electron. Moreover it was found that these electrons had what is called spin half. It's a bit like a spinning top but in a Quantum mechanical sense.

The mystery about the Dirac equation was that apart from the spinning electron there were a number of other solutions with very low energy – in fact negative energy. Negative energy has no meaning in classical physics, but it can be explained away in Quantum theory. Think of it like this: if you are on the ground, you do not have any energy. But if you climb on to a ladder you have positive energy – you can do work. For example like a pile driver, you could jump down and flatten a tin! Now consider the opposite case – you are trapped in a well. Then someone will have to do work to pull you back to the ground. In this sense, you had negative energy, when you were in a well. Similarly also, orbiting electrons in an atom have to be given energy to eject them out. So such electrons have negative energy.

The puzzling question that confronted physicists was, why doesn't the electron plunge down into these lowest energy states – just as a detached apple drops down to the ground without hovering in mid air? That would mean we should not be able to see any electrons in the universe. They would have all plummetted down to negative energies, just as water finds its level. Dirac came forward with a brilliant if crazy, hypothesis to explain this impasse. He said, wait a minute, if a particle is spinning like the electron then two electrons with the same spin cannot be accommodated within the same energy level. It is a bit like two protons not coming together under usual conditions. We know this from another rule in Quantum mechanics, called the Pauli Exclusion Principle. Otherwise this would contradict some experiments. So suppose that all the other negative energies, an infinite sea of energies in fact, are occupied by particles, by other electrons. Then the electrons we see cannot simply plunge down to those energy levels as they are filled up. That is why we see all these electrons in the universe apart from the negative sea which is filled with electrons.

If this crazy possibility be true then something even crazier can happen. Sometimes we should also be alive to the possibility that an electron from the negative sea pops out. Very temporarily no doubt, but it could pop out. Within that wink of time – less than a blink – there is a hole in the negative sea. What sort of a hole is this? Just like an electron except that we are now missing a negative charge. In other words it should appear like a positive charge so that when the original electron would collapse back into the sea, everything would be the same again. Such a hypothetical positively charged particle would be called a positron. It would look exactly like an electron except it would have the opposite charge. Positrons were indeed discovered. This was a great tribute to and triumph of intellectual ingenuity.

In the meanwhile what of the neutrons? It was expected they would decay into a proton and an electron. But this is not quite the whole story. Something didn't add up here. For example, an electron-proton composite would be a boson, but the neutron was a fermion.

The famous physicist Wolfgang Pauli suggested that there would be a third particle along with the electron and the proton which too would be the product of this decay of the neutron. The third particle was called the neutrino. It would be without any charge of course, so that the proton and the electron would neutralize each other and you would have the neutral neutron. The neutrino too was detected though indirectly. In fact a whole zoo of particles as physicists call it has been discovered over the decades. Today we have nearly two hundred and fifty of them. It has become more of a jungle, than a zoo. All these are called elementary particles.

So what exactly is an elementary particle? It's a question like, what is life? The fact is we don't have an exact answer except go with a negative, with the opposite. An elementary particle is a particle which is not a composite. (This is not very satisfactory, as you will realize. It raises the question – is the neutron an elementary particle?). So for instance the hydrogen atom is not an elementary particle but the proton is and separately, so is the electron. In any case we have these two hundred and fifty elementary particles though many of them are very short lived. In just a fraction of a billionth of a second they would disappear. Then how do we know that they exist? Well the answer is, they leave tracks or trails like any good criminal. These are quite literally tracks which can be photographed. In fact if one has a photographic plate these elementary particles would react with the chemicals in the photographic plate and one can see the end products or deposits. These are the tell tale trails.

The elementary particles can also be detected in gas chambers, though not the criminal variety. These are called sometimes bubble chambers and so on. They hold the key to our ultimate understanding of nature, nature in its lowest terms. And what are the characteristics of that smallest scale of nature? These are the very important questions which physicists face.

The positron itself was discovered in the early 1930s independently in the United States and in England. In the US C.D. Anderson spotted it and in England P.M.S. Blackett observed it. The positron is very similar to the electron except that it has a positive charge. There was another interesting feature too. A positron and an electron could form some sort of an atom just as a proton and an electron can. Well, such an atom was discovered in the US by M. Deutsch. Our normal atoms, hydrogen atoms with a proton and an electron are quite stable. But such an atom which was called positronium could live only for about one ten millionth of a second. Very short but still not too short to be observed. Within that time the positron and the electron would collide and destroy or annihilate each other. The resulting energy would appear in the form of what are called gamma rays. These are electromagnetic waves but with a wavelength much smaller than even the x rays. So a positron and an electron annihilate each other and are replaced by two gamma rays or two gamma photons which are moving in opposite directions.

The question now was: The electron had the positron which was an anti particle meaning it was a twin of the electron. It had the same mass and spin, everything except that it had positive charge. Then may be other charged particles too would have anti particles. Further if particles and anti particles collided they would destroy each other with gamma radiation being vented out. So, ordinary particles constitute what we call matter and the anti particles would constitute anti matter. If matter and anti matter were to meet, that would be catastrophic indeed. It would be like hundreds or trillions of hydrogen bombs exploding. If one particle meets an anti particle that's not such a catastrophe though.

You may wonder, why should we live in a universe dominated by matter? It is like saying that males far outnumber females. This is some sort of an asymmetry. We would like to believe that the universe is even handed and doesn't like asymmetry. At least we would like to be able to explain why there is an asymmetry. So some scientists suggested that there would be anti matter as much as matter, but it would be very very far away. Let us say that it would be in an anti universe. Some of these ideas were put forward by a very well known American physicist called Edward Teller who is considered to be one of the fathers of the Hydrogen Bomb. He suggested that matter and anti matter would meet in a catastrophic explosion. This led to an interesting poem that appeared in the New Yorker magazine in

1956. The poem goes something like this [39]:

Well up beyond the tropostrata
There is a region stark and stellar
Where, a streak of anti-matter,
Lived Dr. Edward Anti-Teller....

One morning, idling by the sea,
He spied a tin of monstrous girth
That bore three letters: A,E,C
Out stepped a visitor from Earth.

Then, shouting gladly o'er the sands,
Met two who in their alien ways
Were like as lentils. Their right hands
Clapped, and the rest was gamma rays,

- H.P.F.

What about an anti proton? Could there be a particle exactly like the proton but with the opposite charge? It took a long time to find the anti proton. The first clear evidence for the anti proton came late in 1955 through the work of Chamberlain, E Sagre, C. Wiegand and T.Ypsilantis at the University of California in Berkeley. As expected the mass of the anti proton turned out to be almost the same as that of the proton. In fact the only difference was the opposite charge.

Then of course we have the neutron which has no charge but is otherwise quite similar to the proton. It has almost the same mass and the same spin too. Could a chargeless particle like the neutron also have an anti particle? The answer came in 1956. An anti proton beam that year was made to impact on matter. Certain neutral chargeless particles emerged as a result. These chargeless particles caused further reactions that resembled annihilation. This annihilation was due to particles which had no charge. The conclusion was that anti neutrons had been produced.

You might wonder what anti neutrons are? In what way do they differ from ordinary neutrons? Remember both are chargeless. The answer would be that, neutrons in a sense have some internal structure with charged particles. They decay as you know into protons and electrons and the neutrino. The anti neutrons would have exactly the opposite type of a structure. So you can expect them to break up into positrons, anti protons and anti neutrinos. This established the existence of anti particles.

Meanwhile physicists confronted another puzzle. It's very well to say that there is a proton and an electron in a hydrogen atom. But more complicated atoms had several protons in them – and protons repel each other. Then how do they stay together in the nucleus? The Japanese physicist Hideki Yukawa guessed that perhaps there is some sort of a nuclear force which attracted two protons together, but only at a very very short range. This force vanishes at large distances. Only when two protons come that close does the nuclear force assert itself, overcoming the usual electrical repulsion. It's an attractive force which is not dependent on the charge and so binds the protons together. Yukawa, based on such ideas predicted in the mid 1930s that there would be some sort of a particle which would be a glue, carrying this nuclear force. He even suggested that such a particle would have a positive charge so that even a neutron can be described as being bound to the nucleus.

You could picture the whole thing this way. A neutron acquires one of these new Yukawa positively charged particles and gets transformed into a proton or a proton would give up this particle and get transformed into a neutron. Identities at this level of analysis keep switching. Later it was suggested that there would be even negative particles of the Yukawa type, and why not then neutral ones?

In 1947 a group of physicists in England namely C. Lattes and coworkers observed a particle track in nuclear emulsions. This track seemed to indicate exactly a Yukawa particle. Moreover it came with positive, negative and neutral charges. This was christened by them as the pi-meson. Mesons, because they were "middle" particles – not as heavy as protons, and not as light as electrons. So, slowly, partly through theory and partly through observation, the number of elementary particles was filling up.

Around this time when hardly six or seven elementary particles were known, Y. Nambu who won the Nobel Prize in 2008, put forward a very simple formula to give all their masses. It was pure guess, an ad hoc formula. Take an integer Nambu said, add half to it. Such numbers are called half integers as you may know. Now multiply this half integer to the mass of a pi-meson (or pion for short). For six different integers you will get the masses of all the known six elementary particles.

An important question is, why are the properties of elementary particles so important? This is because if we are to understand nature fully, we have to come down to its smallest constituents. This is really what atoms were all about and the atomic philosophy has been extended now to elementary particles. This approach is called reductionism – understand the large in terms of its smallest building blocks. So by knowing the building blocks, the elementary particles, and how they interact with each other, in other words what are the forces which make them behave as they do, we can understand the very fundamentals of nature. That was the whole game.

Another question you may ask is: How do we study their properties? Because they are too small to be seen even with the most powerful instruments and some of them live barely for the minutest fraction of a second.

Protons and electrons which make up the hydrogen atom are the only known stable elementary particles. The neutron itself as we saw decays after several minutes. It is not strictly speaking, stable. When we come to other elementary particles the situation is worse. We are in a regime where one millionth of a second is a long life time. We cannot in any way study these elementary particles directly. It is still possible to trap some of these particles for a long time. But we would need a huge amount of energy to make this possible. One gram of elementary particles to be trapped in a small container would require the amount of electricity which a big country consumes. That is preposterous!

So we clearly have to use other methods. What can we do? Firstly, we produce protons or electrons in nuclear or even sub nuclear reactions. Then we have a beam of these particles go and collide with another such beam of particles or even a target which contains these particles. We then have to observe what happens during these collisions. Incredible as it may seem the ultimate secrets of the universe tumble out of the by products of the results of these collisions. But how do we "see" these results? Remember, elementary particles after they are produced, travel very fast and decay very fast. There are different ways in which we could indirectly detect what happens in these collisions.

For instance, when these elementary particles strike other atoms they excite these atoms. The electrons orbiting the nuclei in these atoms acquire energy and thereby they could be completely ejected from the atoms with this acquired energy. Then the atoms would be ionized. Or they could jump into a state of higher energy within the atom. Then of course they crash back to their original stable energy. In the process this acquired energy is emitted as some sort of electromagnetic radiation, generally light. Even in ionization, light can be emitted. By observing this type of emitted light, we can extract some information about the particles that have been produced, and where and how. This in turn would give us some idea about the forces acting between these elementary particles. Ingenuity to beat the best of Sherlock Holmes!

Alternatively, we could use special sensitive photographic emulsions through which the final products of the collision are made to traverse, as noted a little earlier. The charged, rapidly rushing particles, react with the silver bromide present in the photographic emulsion and after proper chemical treatment these regions of interaction show up as trails of silver grains. These tracks are the footprints of the elementary particles that have been created by the collision.

Yet another method that was popular some decades ago was the use of what is called a cloud chamber. This was built by C.R.T. Wilson in Edinburgh. As the name suggests it resembles the recreation of a cloud. A chamber is filled with air and some vapor, of an alcohol for example. The vapor is completely saturated so that a little cooling will cause condensation. Such a sudden cooling is provided to the chamber and condensation can then take place even on ions inside the chamber. The elementary particles leave a track of ions and condensation takes place on these ions and we see a trail of cloud.

Then there is the bubble chamber. This is based on the discovery made by D. Glaser in 1952 in the United States. In this case a liquid is kept under a certain pressure. If the pressure is suddenly released then the liquid begins to boil and bubbles stream out, for example along the trail of charged particles which is left behind.

There are other well known techniques too like the Geiger counter, the scintillation counter and so on which tick when struck by a particle. An interesting technique is that of the Cherenkov counter. This is based on a discovery made by a Russian physicist. He noticed that if a particle travels through a transparent substance with a velocity greater than that of light in that substance, then the electrons and the molecules of the substance release a sort of a weak glow which is called Cherenkov radiation. So a particle can leave a trail through a transparent medium in the form of the Cherenkov radiation. These are the footprints of the particle itself.

The question you may ask is, all this is fine but what do we do if the particle does not have a charge? Well they are detected indirectly by the means of charged particles that are produced by these neutral particles in different ways. To get an idea of how this works, consider an experiment to detect neutrons. They normally do not leave any track in the nuclear emulsion because they are neutral. But what happens when these neutrons collide with the hydrogen atoms present in the emulsion? Well the neutrons can occasionally knock out the protons in the hydrogen and these protons as they pass through the emulsion could leave tracks behind. This is an example of an indirect method of tracking the particles.

5.2   Fundamental Forces

It is by studying the most fundamental or elementary particles that we can
hope to understand the most fundamental aspects of nature. The interactions or forces between the elementary particles would provide us with the
ultimate information about the forces of nature. As we have seen there are
three main forces between these particles. The most obvious is of course the
electromagnetic force which is attractive for opposite charges and repulsive
for like charges. There are two other less obvious interactions amongst elementary particles. We have encountered these too. One is the interaction
which causes decay – for example the neutron decaying into a proton, an
electron and a neutrino. This force is called the weak force – it is weaker
than the electromagnetic force. The third force between elementary particles is the so called strong force that holds two protons together within
the nucleus, whereas they should really be repelling each other. As we saw,
we required this force to be much stronger than the electromagnetic force,
though only at very small distances, and always attractive.

Even if we could understand in detail each of these interactions, that would
not be the end of the story. The question that would come up is, are these
fores really different or do they share anything in common? In other words
we would like as a first step at least, to give a unified description of these
interactions.

Finally there is of course the gravitational interaction. As Newton had
observed, every bit of matter, every particle including the elementary particles would attract other elementary particles with this gravitational force.
Though ultimately gravitation is also important if we are looking for a unified description of everything, in practical terms the gravitational force is
so much weaker than the other forces that it is all but negligible. It is
less than a trillion trillion trillionth of the electromagnetic force. However
gravitation dominates in the large scale universe. As almost all atoms are
neutral – the negative charges and positive charges cancel each other – this
helps the cause of gravitation. In any case, it has not, as yet been possible
to give a unified description that includes gravitation.

If we look back at prehistory, we find bewildered man assigning to different
natural phenomena, different controlling powers or deities. But gradually,
we could discern underlying common denominators. Over the millennia
man's quest for an understanding of the universe has been to explain apparently disparate phenomena in terms of a minimal set of simple principles.
Today with hindsight we can see the logic of Occam's razor (literally, "A

satisfactory proposition should contain no unnecessary complications"), or an economy of hypothesis [40]. This has been a sacred guideline for physicists. This is indeed a far cry from prehistoric times.

In the words of a great American physicist F.J. Dyson, ".... the very greatest scientists in each discipline are unifiers. This is especially true in Physics. Newton and Einstein were supreme as unifiers. The great triumphs of Physics have been triumphs of unification. We almost take it for granted that the road of progress in Physics will be a wider and wider unification...".

Sir Isaac Newton was the first great unifier. He discovered the Universal Law of Gravitation: Earlier, as we saw, Kepler had summarized the meticulous planetary observations of Tycho Brahe, in three laws. These were equivalent to the single law of gravitation. Moreover the force which kept the moon going round the earth, or the earth round the sun was also the force which kept binary stars going around each other. And so on. All this was basically the same force of gravitation which brought apples down from a tree. This apart his Laws of Motion were also universal.

In the nineteenth century the work of Faraday, Ampere and others showed the close connection between the apparently totally dissimilar forces of electricity and magnetism. It was Maxwell as we saw, who unified electricity not just with magnetism but with optics as well.

There was another great unification in the nineteenth century: Thermodynamics linked the study of heat to the theory of motion of molecules in gases.

In the early part of the twentieth century Einstein fused space and time, giving two supposedly separate entities an inseparable identity, spacetime. As we discovered, he went on to unify spacetime with gravitation in his General Theory of Relativity. However the unification of electromagnetism and gravitation has eluded several generations of physicists, Einstein included.

Meanwhile, we saw that thanks to the work of De Broglie and others, the newly born Quantum theory unified the two apparently irreconcilable concepts of Newton's "particles" and Huygen's waves. This unification of two long recognized but distinct characteristics as being different sides of the same coin was indeed the bedrock of Quantum theory as noted earlier.

Another unification in the last century is the fusion of Quantum mechanics and special relativity by Dirac, through his celebrated equation of the electron. This unification is not so obvious, but it lead to the discovery of so many properties and the anti-particles as we already noted.

A very important unification took place in the sixties and seventies due to the work of the Pakistani British physicist, Abdus Salam, and two other American physicists, Steven Weinberg, and Sheldon Glashow and others– the unification of electromagnetism with the weak forces. This has given a new impetus to attempts for unifying all interactions, gravitation included. The weak force is one of two forces, the other being the strong force, discovered during the twentieth century itself. As we saw, earlier studies and work revealed that there seemed to be three basic particles in the universe, the protons, the neutrons and the electrons. All atoms could be built out of these three ingredients. While the proton and the electron interact via the electromagnetic force, in the absence of this force the proton and the neutron appear to be twins. If you ignore the charge, it is very difficult to distinguish a neutron from the proton, as they have the same spin and almost the same mass. In the atomic nucleus the proton and the neutral neutron interact via the same "strong forces", which bind protons and protons. These forces are about ten times stronger than the electromagnetic but beyond a thousand billionth of a centimeter, they disappear, and electromagnetism then dominates.

The existence of the neutrino was postulated by Pauli in 1930 to explain the decay of the neutron. This particle was discovered by Reines and Cowan in 1955. The weak force which is about a thousand billionth the strength of the electromagnetic force is associated with neutrino type particles and has an even shorter range, less than a thousandth that of the short range strong forces. The neutrino itself has turned out to be one of the most enigmatic of particles, with peculiar characteristics, the most important of which is its handedness.

To understand this let us board our Space Lab and zoom off to some distant inhabited planet. To our surprise we find that all the people there are ambidextrous. They are equally at ease being lefthanded or righthanded including in writing. In a sense, all elementary particles are like that. Except the neutrino. This is like a species on the alien planet that is only lefthanded. In the context of the neutrino, what does this mean? Consider a spinning neutrino (Cf.Fig.5.1). We can classify this as lefthanded, if we can imagine the fingers of our left hand wrapped round the neutrino in the direction of the spin. The thumb points upwards and that is the direction in which it moves. Strange as it may seem, you cannot make a neutrino go down without changing its direction of spin! That is reserved for the antineutrino. The anti-neutrons are all righthanded. This handedness property appears to be crucial for weak forces.

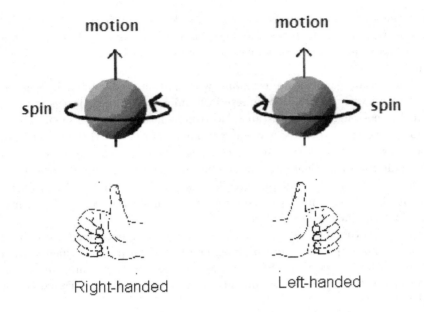

Fig. 5.1   Neutrino Spin

Later work revealed that while particles like the electron and neutrino, what are called light particles or leptons may be "truly" elementary, particles like the protons may be composite, in fact made up of still smaller constituents called quarks – six in all. Today it is believed that the quarks interact via the strong forces. If neutrinos are queer, quarks are stranger still. They are the only particles which break the rule that all charged particles have charges which are integers in units of the electron or proton charge. Quarks have charge one third or two thirds (plus or minus)! Moreover, they too display neutrino type handedness. Finally, though the idea of quarks is over four decades old, none have been observed todate. Worse, physicists assure us that no free quarks can ever be seen. Then how can we believe in their existence, you might ask. Scientists believed in atoms long long before they were actually observed. The point is, if we postulate quarks, then there should be several properties, several footprints. All these have been confirmed. The important thing is, we can now explain the strong forces, which if you remember are always attractive and strong enough to overcome the repulsion of protons in a nucleus. Now we can say that the strong force acts between the quarks that make up proton A and proton B, binding them together. In this sense, the strong force does not act between proton A as a whole and proton B as a whole, as was believed earlier.

All "material" particles are fermions, with half integral spin. It is believed that forces or interactions originate in fermions. But these forces are carried by messengers. For instance electromagnetic forces are mediated by photons which are bosons, with integral spin, spin 1 in fact. This is crucial, for a mathematical formalism called gauge theory which can describe all these interactions. There is, though one exception. The gravitational force is carried by a particle of spin 2, called the graviton. That is the problem: gravitation, though its inverse square law resembles the electric force, just does not gel with the other forces.

To picturize the above let us consider the interaction between a proton and an electron. A proton could be imagined to emit a photon which is then absorbed by the electron. Such studies, in the late forties and fifties culminated in the highly successful theory of Quantum Electro Dynamics or QED.

Instead of a single mediating particle we could even think of multiplets, all having equal masses. You may remember the family of three Yukawa mesons.

So the following scenario emerges: When we say that a proton A and an electron B (or even a proton) interact, we mean that A ejects a photon which reaches B and vice versa. All this happens within a wink. You have to keep in mind that the photons carry energy, and there is a Quantum mechanical uncertainty in time, just as there is a position momentum uncertainty. We can't really detect anything within this short uncertain interval of time. In other words the photons are virtual.

Several physicists in the middle of the last century conjectured that the weak interaction too would be mediated by spin 1 bosons. It was realized that three such particles would be required, two W particles with opposite charge and a chargeless Z^0 particle. Unlike the massless photon however, this triplet of particles would be massive. The reason is that a massless particle like the photon travels with the speed of light and can go a long long way. That is why the electromagnetic interaction is long range. The weak interactions however as we saw are short range and so have to be mediated by heavy bosons which therefore travel much shorter distances with speeds lower than that of light.

The question was how could the photon be massless while the W and Z

particles would be massive? [42]. And how does all this give a unified description? To appreciate this, let us pick up a bar magnet. It has a North Pole and a South Pole. Let us now heat it. It looses its magnetism. In effect the North and South pole asymmetry is broken. Conversely, when the semi-liquid magnet cools down, polarity or asymmetry is restored spontaneously. You don't have to make this happen. This in fact is a transition from symmetry to asymmetry. Physicists call it a phase transition.

In our case, at a very high temperature there is total symmetry: the W and Z particles as also the photon are all indistinguishable. The Ws, Zs, and the photons would all be massless. After the phase transition, while the photons remain mass less, the others would acquire mass. A transition would occur at temperatures like a thousand trillion degrees Centigrade. At even higher temperatures there would be a single "electroweak" force. As the temperature falls to the above level electromagnetism and weak forces would separate out much like the North magnetic pole and South magnetic pole. This scheme was as noted worked out in detail independently by Abdus Salam and two Americans, Sheldon Glashow and Steven Weinberg. They still had to explain how the massless W and Z bosons became massive. They invoked a process suggested by superconductivity theory – a massive boson would be required for generating the mass. (This was christened the Higgs boson, after Peter Higgs of England who had suggested this mechanism in the mid sixties). The trio got the Nobel Prize for this achievement.

Next how do we include the strong forces? Clearly the direction to proceed appeared to be to mimic the same game plan. The strong force would be mediated by spin one particles, the gluons. (The approach differed from the earlier version of strong interaction in terms of Yukawa's pions as you might have noticed.) This force binds the different quarks to produce the different elementary particles, other than the leptons (electrons and neutrinos) which are elementary in themselves. This is the standard model, the twentieth century peak of achievement of Quantum theory. It must be mentioned that in the standard model, the neutrino is a massless particle. However we have not yet conclusively achieved a unification of the electroweak force and the strong force. We proceed by the analogy of the electroweak unification to obtain a new gauge force that has been called by Jogesh Pati and Abdus Salam as the electro nuclear force, or in a similar scheme the Grand Unified Force by Glashow and Georgi. It must be mentioned that one of the predictions is that the supposedly stable proton would decay, though with a life time much much more than the age of the universe itself. It may still be possible to detect the decay of the proton, without waiting for the end of the universe. This is because, the life times are really "half lifes" – a span of time in which a certain proportion of the particles would decay. Some even believe that we are near a situation where this should be observable.

So what we have seen is that three ingredients have gone into this marvel of the standard model. The first is that all the forces have been described in terms of suitably chosen bosons, all having the spin one. The second is that elementary particles like protons, mesons and so on, excepting electrons and neutrinos, are all composites of quarks, interacting via the strong force. Finally there is this mechanism of spontaneous breakdown of symmetry which distinguishes the various sets of interaction bosons. (Prof. Salam had another analogy for this breakdown of symmetry. Consider a well laid out circular dinner table. Every diner has a spoon, fork and knife to his left and right, just before service begins. There after one diner picks up the fork and spoon to his left – and so will everyone else have to. The symmetry is broken!).

This "unifying" theory still relies on eighteen parameters whose ad hoc values are expected to explain observation, rather than follow from any fundamental principle. The standard model is also plagued by problems like the "hierarchy problem". This arises from the widely different energies or masses that have to be associated with the various interactions. There are also the as yet undetected monopole, and infinite quantities or divergences (which have to be eliminated by mathematical trickery). And so on. A decade ago the Super Kamiokande detector in Japan detected a neutrino mass. This is the first evidence of what may be called, physics beyond the standard model. Remember the neutrino is supposed to be massless. Interestingly this means that now we also require a right handed neutrino. Here is how the Dutch Nobel Laureate, Gerard 't Hooft (drawing a comparison with planetary orbits) describes the inadequacy of the standard model [43], "What we do know is that the standard model, as it stands today, cannot be entirely correct, in spite of the fact that the interactions stay weak at ultrashort distance scales. Weakness of the interactions at short distances is not enough; we also insist that there be a certain amount of stability. Let us use the metaphor of the planets in their orbits once again. We insisted that, during extremely short time intervals, the effects of the forces acting on the planets have hardly any effect on their velocities, so that they move approximately in straight lines. In our present theories, it is as if at short time intervals several extremely strong forces act on the planets, but, for some reason, they all but balance out. The net force is so weak that only after long time intervals, days, weeks, months, the velocity change of the planets become apparent. In such a situation, however, a reason must be found as to why the forces at short time scales balance out. The way things are for the elementary particles, at present, is that the forces balance out just by accident. It would be an inexplicable accident, and as no other examples of such accidents are known in Nature, at least not of this magnitude, it is reasonable to suspect that the true short distance structure is not exactly as described in the standard model, but that there are more particles and forces involved, whose nature is as yet unclear."

5.3 Quantum Gravity

The theory of relativity (special and general) and Quantum theory have been often described as the two pillars of twentieth century physics. Each in its own right explained aspects of the universe to a certain extent. For example, as we have just now seen, the standard model goes a long way in explaining the strong, weak and electromagnetic forces. But there are still many unanswered questions. For instance spacetime singularities (like the Big Bang), designated by John Wheeler as the "Greatest Crisis of Physics", the many infinite quantities encountered in particle physics, or as we have seen some eighteen arbitrary parameters or values in the standard model, several as yet undetected entities like the elusive monopoles and Higgs bosons, gravitational waves and dark matter and so on and so forth. I got a petulant letter from a British computer scientist a few years ago. Why do not physicists begin their papers listing all the things that have not yet been found, even after decades, he asked.

All this apart, it was almost as if Rudyard Kipling's "The twain shall never meet" was true for these two intellectual achievements - general relativity or gravitation and Quantum theory, a view endorsed by Pauli, who went as far as to say that we should not try to put together what God had intended to be separate. For decades there have been fruitless attempts to unify electromagnetism and gravitation, or Quantum theory and general relativity: As Wheeler put it, the problem has been, how to incorporate general relativistic spacetime curvature into Quantum theory or spin half into general relativity [33]:

"It is impossible to accept any description of elementary particles that does not have a place for spin $\frac{1}{2}$. What, then, has any purely geometric (i.e. general relativistic) description to offer in explanation of spin $\frac{1}{2}$ in general? More particularly and more importantly, what place is there in quantum geometrodynamics (i.e. a general relativistic description) for the neutrino–the only entity of half-integral spin that is a pure field in its own right, in the sense that it has zero rest mass and moves with the speed of light? No clear or satisfactory answer is known to this question today. Unless and until an answer is forthcoming, pure geometrodynamics must be judged deficient as a basis for elementary particle physics.

We must admit that decades of attempts to provide a unified description of these two theories have failed. Einstein himself lamented, in later years, "I have become a lonely chap who is mainly known because he doesn't wear socks and who is exhibited as a curiosity on special occasions."

Einstein's lament had a long history. Around 1915 when he formulated his general theory of relativity he had already realized that it would be essential to have a unified description of gravitation and electromagnetism, the only two known forces at that time. Soon enough there were two novel attempts. One was by the great German mathematician and physicist Hermann Weyl. He introduced what was called Gauge Geometry. The idea was revolutionary. Just as lengths and intervals of time could vary with the speed of the observer, Weyl proposed these intervals and lengths could also depend on the location in space. In other words even if two observers were not moving with respect to one another they would still measure different lengths, for the same rod, purely by virtue of the fact that they were at different points in space. However novel the idea was, Einstein could not accept it as a real unification of the two forces. He pointed out that there was the ad hoc feature of electromagnetism being inserted into the theory by force, rather than come up as a natural feature of spacetime.

Another attempt, equally novel was made by a young German, Theodore von Kaluza. He demonstrated that we could get gravitation and electromagnetism from the same equations, but there was a price to pay. These equations had to be formulated in five dimensions, not our usual four dimensions of real life. This idea was developed by the Swedish mathematician Oskar Klein. Of course the fifth and extra dimension, had to be explained. Why do we not experience it? The answer was that the fifth dimension, unlike the other four was miniscule, so small that we could not perceive it in real life. We can easily understand how it works. Think of a cigarette. Now let the same amount of tobacco be wrapped in a cigarette that is hundred times as long. What would happen is that the cigarette, which was in a cylindrical shape to start with, would now become a very very thin cylinder. If the cigarette were made longer and longer, the cylinder would practically disappear into a one dimensional straight line – almost. The extra dimension of thickness would all but vanish.

In any case Kaluza sent the paper to Einstein and interestingly there are two versions of what happened thereafter. According to Abdus Salam, Einstein sat on the paper for a long time. Kaluza was so discouraged that he gave up physics and took to the theory of swimming. According to another version however, Einstein was actually thrilled by this idea and even discussed this possibility of unifying electromagnetism and gravitation with other physicists. Other attempts too were made introducing extra dimensions of space. All these were unsatisfactory. But bear in mind that this was an attempt to unify the two forces, and not the unification of Quantum theory and general relativity.

Heisenberg and Pauli the celebrated physicists were amongst the first to attempt such a unification around 1930. Their motivation was that the methods of Quantum Field Theory could work here, in particular the successful Quantum mechanical treatment of electromagnetism. This treatment is based on a technique that can be called the quantization of a field. In the case of electromagnetism, which is a field spread all over space, what this technique does is, it sort of chops up the field into tiny discrete bits. These bits are the photons, the very same photons that Max Planck had introduced much earlier. Couldn't the same technique be applied to the gravitation field which is also spread through space? This time we would get gravitons instead of photons.

Unfortunately however the technique fails in the case of gravitation. This is because, in the case of electromagnetism, we encounter infinite quantities which can be bypassed using mathematical devices. In the case of gravitation too we get infinite quantities. Unluckily however this time round, due to the nature of the gravitational field and its equations, the infinite quantities cannot be bypassed. The reason for this can be thought of like this. The gravitational field has energy – and energy means some sort of mass. Now mass means gravitation again! We saw this earlier. In other words there is a cascading self build up. This complication is not there for electromagnetism. This has remained an unsolved problem for over seventy five years.

At the same time it is also remarkable that both these disparate theories, Quantum mechanics and general relativity, share one common platform:

In both these theories, space and time, or spacetime are smooth, with no kinks or holes. It is rather like a smooth silk fabric viewed from afar. So scientists, pushed to the wall, have renounced this time honoured prescription. This is rather like Kepler finally discarding the Greek circular orbits. Revolutions are generally forced upon us.

Smooth spacetime has been questioned by what are called Quantum Gravity theories including my own fuzzy spacetime model on the one hand and Quantum superstrings on the other amongst more recent approaches. These try to provide the elusive unified description.

While we will return to these approaches, let us first take a small detour and explore briefly some rather surprising ideas about "non-smooth" space. Here we begin to jettison some once sacred concepts. Fresh air is coming into the new schools of thought. The story begins in 1867, when Kirkwood, an American astronomer made a startling discovery about the asteroid belt. The asteroid belt is the teeming corridor of the thousands of rock like objects, swirling between the orbits of the planets Mars and Jupiter. These objects were not uniformly distributed - there were gaps. Moreover, these gaps were in regions where the periods of revolution of the asteroids have a simple ratio to that of Jupiter. This discovery vexed astronomers for over a century till Wisdom, with the help of modern computers showed that in these specific regions and at these specific revolution periods, the orbits become chaotic. This means that they become very sensitive to small changes and so they are unpredictable [44]. This means, in the remote past, the asteroids were uniformly distributed as we expect. But the asteorids present in these chaotic regions would have over time, been ejected out.

So the nineteenth century saw the birth of doubts about the stability of orbits in the solar system, an issue that had apparently been settled in the context of the rigid determinism of Newton and Laplace. By the turn of the twentieth century, the question had been, in principle, answered by Poincaré. Not just the solar system with its thousands of objects, Poincaré as we saw, showed that even a simple system of three objects would be unpredictable. As he noted [45], "If we knew exactly the laws of nature and the situation of the universe at the initial moment, we could predict exactly the situation of that same universe at a succeeding moment. But even if it were the case that the natural laws had no longer any secret for us, we could still know the situation approximately. If that enabled us to predict the succeeding situation with the same approximation, that is all we require, and we should say that the phenomenon had been predicted, that it is governed by the laws. But it is not always so; it may happen that small differences in the initial conditions produce very great ones in the final phenomena. A small error in the former will produce an enormous error in the latter. Prediction becomes impossible." In other words, minute changes here and now, could lead to major effects there and then. Quite contra to our intuition [46]. Indeed, this is the type of thinking behind the famous butterfly effect – how the flapping of wings of butterflies in South America, as an example, could lead to major weather changes in the African continent. In any case, this is more of a metaphor.

Prigogine begins the book, "Exploring Complexity" [47] with the following statement, "Our physical world is no longer symbolized by the stable and periodic planetary motions that are at the heart of classical mechanics. It is a world of instabilities and fluctuations..."

In fact scholars have since demonstrated this chaotic behaviour of the supposedly stable solar system orbits. The stability is approximate, being valid over relatively short periods of time. Short periods could be from centuries to millions of years.

Let us now come to a more down to earth problem that vexed Richardson [48]. He found that the length of the common land boundaries claimed by Portugal and Spain as also Netherlands and Belgium, differed by as much as twenty percent! A grave border dispute, except nobody noticed it! The answer to this non-existent dispute lies in the fact that we are carrying over our concepts of smooth curves to the measurement of real life jagged boundaries or coastlines. As the French-American mathematician Mandelbrot puts it "The result is most peculiar; coastline length turns out to be an elusive notion that slips between the fingers of one who wants to grasp it. All measurement methods ultimately lead to the conclusion that the typical coastline's length is very large and so ill determined that it is best considered infinite....."

What this means is that if you measure the length using a meter as your unit, you will get one answer. If you use centimeters, you will get a different length! Infact smooth curves are a mathematician's idealization. In the words of Dyson [49] "Classical mathematics had its roots in the regular geometric structures of Euclid and the continuing dynamic structure of Newton. Modern mathematics began with Cantor's Set Theory and Peano's Set-Filling Curve...The same pathological structures that the mathematicians invented to break loose from 19th Century naturalism turn out to be inherent in familiar objects all around us". Dyson was referring to the work of the great nineteenth century mathematician Georg Cantor. To picturize a Cantor set, think of a continuous straight line. Now remove any number of points from this line, so that there are as many holes as points, totally intertwined. You can do this by, for instance, first deleting a third of the line from the middle. Next delete a third of a remaining segement from its middle. And go on and on. The line now is almost continuous, but no place in it is really continous.

The advent of Quantum mechanics at the turn of the twentieth century brought into physics many strange concepts like the Uncertainty Principle, complex wave functions, and in general, as in the coastline situation, an observer dependent universe. These radical ideas have troubled some of the greatest minds of the twentieth century - Einstein for example – and continue to do so. But, and this is a point we will often stress, the new Quantum theory was built on the foundations of the ruins of classical theory, in the sense that not only are the Quantum mechanical equations deterministic but they are set in a continuous space and time, rooted as they are in concepts like spacetime points and point particles. Is this at the root of some of the shortcomings of Quantum theory, which we briefly touched upon in the preceding chapter and will encounter again?

The truth is, in the real world, chaos (in a literal, rather than the narrow mathematical, technical sense) and randomness are natural while order is artificial and constitutes special cases. It is more an idealization of the human mind. Mandelbrot [50] quotes French Nobel Laureate Jean Perrin, "At first sight the consideration of the general case seems merely an intellectual exercise, ingenious but artificial, the desire for absolute accuracy carried to a ridiculous length. Those who hear of (non smooth) curves without tangents, ... often think at first that Nature presents no such complications, nor even suggests them....

"The contrary, however, is true, and the logic of the mathematicians (in considering "pathological" structures like non-differentiable non-smooth curves) has kept them nearer to reality than the practical representations employed by physicists. This assertion may be illustrated by considering certain experimental data without preconception...

"We are still in the realm of experimental reality when we observe under the microscope the Brownian motion agitating a small particle suspended in a fluid. The direction of the straight line joining the positions occupied at two instants very close in time is found to vary absolutely irregularly as the time between the two instants is decreased...

"It must be borne in mind that, although closer observation of any object generally leads to the discovery of a highly irregular structure, we often can with advantage approximate its properties by continuous functions".

The question we will explore in the following lines though from different perspectives, is, "Is the smooth space time continuum an approximation a la the measurement of the length of a jagged coastline by thick brush strokes which conceal the irregularities?" Two approaches which have captured attention are those of string theory, and to a lesser extent, what has been dubbed, "Loop Quantum Gravity". My own approach, which also we will encounter broadly shares this platform, though with important differences.

It is in the spirit of Einstein's, "I want to know how God created this world. I am not interested in this or that phenomenon, in the spectrum of this or that element. I want to know His thoughts, the rest are details."

As Prigogine puts it [51], "At the end of this (twentieth) century, it is often asked what the future of science may be. For some, such as Stephen W Hawking in his Brief History of Time, we are close to the end, the moment when we shall be able to read the "mind of God." In contrast, we believe that we are actually at the beginning of a new scientific era. We are observing the birth of a science that is no longer limited to idealized and simplified situations but reflects the complexity of the real world...."

Indeed Einstein himself had anticipated this. As he observed around 1930 itself "... It has been pointed out that the introduction of a space-time continuum may be considered as contrary to nature in view of the molecular structure of everything which happens on a small scale. It is maintained that perhaps the success of the Heisenberg method points to a purely algebraic method of description of nature that is to the elimination of continuous functions from physics. Then however, we must also give up, by principle the space-time continuum. It is not unimaginable that human ingenuity will some day find methods which will make it possible to proceed along such a path. At present however, such a program looks like an attempt to breathe in empty space."

Infact the great British mathematician Clifford had also anticipated such ideas in the nineteenth century itself, "I hold in fact (1) That small portions of space are in fact of a nature analogous to little hills on a surface which is on the average flat; namely, that the ordinary laws of geometry are not valid in them. (2) That this property of being curved or distorted is continually being passed on from one portion of space to another after the manner of a wave..."

Perhaps smooth spacetime is an approximation? Infact Mandelbroit's work on non smooth objects or fractals has clearly brought out that the smooth

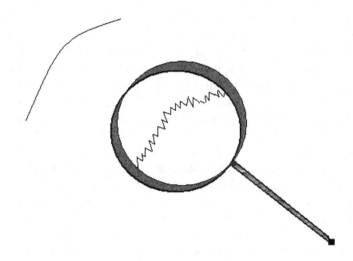

Fig. 5.2   How smooth is the curve?

curves of classical mathematics are to be replaced in real life, by jagged structures, previously dismissed as pathological cases (Cf.Fig.5.2).

At the same time this new description has many ramifications and leads, in my own formulation, to a cosmology which contrary to belief correctly predicted the latest iconoclastic observations, for example that the universe is accelerating and expanding for ever with a small cosmological constant while supposedly sacrosanct constants like the gravitational constant seem to be changing with time. We will see this briefly later on.

5.4   Strings and Loops

From Galilean-Newtonian mechanics to Quantum Field Theory, the concept is Newtonian, in that spacetime is a container or stage within which the actors of matter, energy and interactions play their parts, even modifying the stage. However, the new concept of spacetime is Liebnitzian, in that, the actors create or define the stage itself. This was the view of Newton's German contemporary, Liebniz [52]. The actors *are* the stage. It now becomes possible to circumvent the otherwise insurmountable spacetime singularities and even the famous infinities.

We have noted that in spite of some success, the standard theory has failed to rope in gravitation. Indeed one of the main troublemakers has been the point spacetime of the earlier theories. An easy way to appreciate this is to remember that Quantum theory decrees that if we go down to space or time points, we have to encounter infinite momenta or energies. For the past few decades new approaches have tried to break out of this limitation. Let us first consider string theory.

An Italian physicist, T. Regge in the fifties mathematically analyzed some shortlived particles, called particle resonances. These resonances seemed to fall along a straight line plot, with their angular momentum being proportional to the square of the mass (Cf.Fig.5.3)

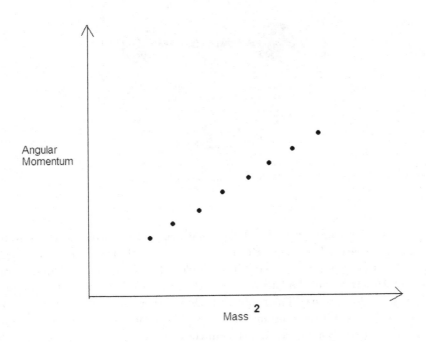

Fig. 5.3   Regge resonances

Surprisingly, this is what you would obtain if you hurl a stick up. The stick rotates as it hurtles on, while the mass and spin follow this type of a graph. All this suggested that resonances had angular momentum, on the one hand and resembled extended objects, that is particles smeared out in space. Remember, we had pictured particles as being made up of quarks. A pi-meson for example consists of a quark-antiquark pair. To explain the Regge plots, you would have to picture a quark and antiquark tied to the ends of a stick – or string, a distance less than a thousand billionth of a centimeter [53]. (Cf.Fig.5.4).

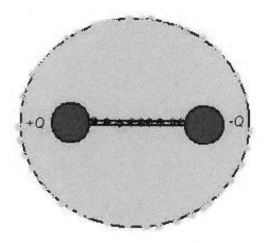

Fig. 5.4   A simple pion model

This went contrary to the belief that truly elementary particles were points in space. In fact at the turn of the twentieth century, Poincare, Lorentz and others had toyed with the idea that the electron had a finite extension. But they had to abandon this approach because of a conflict with special relativity. The problem is that if there is a finite extension for the electron then forces on different parts of the electron would blow it up. In this context, it is interesting that in the early 1960s, Dirac came up with an imaginative picture of the electron, not so much as a point particle, but rather a tiny membrane or bubble. Further, he suggested that the higher energy levels of this membrane would represent the heavier electron type particles like muons [54]. But this too lead to all sorts of problems.

Then, in 1968, G. Veneziano another Italian working at CERN, Geneva, came up with a unified description of the Regge resonances and other processes in which particles collide and scatter. Veneziano considered the collision and scattering process as a black box (Cf.Fig.5.5). He pointed out that there were in essence, two separate channels or modes for scattering. These, he argued gave a dual or double description of the same process.

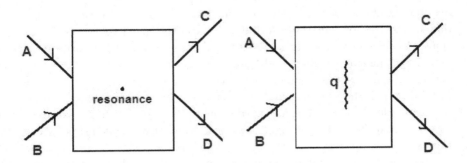

Fig. 5.5   A scattering process from two different perspectives

In one of the channels or modes, particles A and B collide, form a resonance R, which quickly disintegrates into particles C and D. On the other hand we have in a different mode or channel scattering particles A and B approach each other, and rather than form a resonance, they interact via the exchange of a particle q. The result of the interaction is that particles C and D emerge. If we now enclose the resonance and the exchange particle q in an imaginary black box, it will be seen that both the processes describe the same input and the same output: They are essentially the same.

These ideas gelled with an interesting picture of how quarks interact with each other inside an elementary particle that they make up. When the quarks come very close to each other, there is no force acting on them! They move freely, unfettered. But let them try to escape – then they get bound up by the "string" to which they are tethered, as we saw. It is as if the quarks are bound to the ends of a rubber band. They are quite free if the band is loose, but they get bound when the band is stretched. And remember, all this drama takes place within the very tiny confines of less than a million millionth of a centimeter! That is how some of the ideas of particle forces ended up with the tiniest of rubber band like strings.

This description which emerged from the physics of point particles may be called a "Bosonic String" theory, as it gives a description of bosons, but not spinning fermions. These theories soon displayed problems. For example they allow the existence of tachyons which are particles rushing along at speeds greater than that of light [55]. Further they do not easily meet the usual requirements of Quantum theory. That is because, usual Quantum theory is about point particles and points of smooth spacetime. But here we are talking about particles that are somehow smeared or spread out. The difficulties are however resolved only if spacetime had twenty six dimensions. As we saw a little earlier, it is possible to get results by introducing new dimensions a la Kaluza and Klein.

If the string is given rotation, then we get back the Regge trajectories given in (Cf.Fig.5.3) above as we would expect. That is we can include fermions too in the description. Here we are dealing with objects of finite extension rotating with the velocity of light. This lead to what was called super string theory.

Happily the theory of Quantum superstrings requires only ten dimensions. So, is there an explanation for the extra dimensions that appear in string theories? Can they reduce to the four dimensions of the physical spacetime? Physicists invoked the trick suggested by Kaluza and later Klein in the early twentieth century. Kaluza's original motivation if you remember had been to unify electromagnetism and gravitation by introducing a fifth coordinate but make this extra dimension nearly zero by allowing it to curl up.

A finite extension for an elementary particle, as in string theories can be shown to lead to a new type of geometry for spacetime – this is called non commutative geometry [56] In this case two space coordinates like x and y do not commute. What does this mean? Remember that x and y are numbers, and as we have been learning from primary school x times y is the same as y times x just as three times four equals four times three. That is, they commute. But in the new geometry, this is no longer true, strange as it may seem. Let us see what this means. If you go from a starting point 0, a distance x meters along a certain direction to reach a point X, then you turn left and go a distance y meters, you come to a point P (Cf.Fig.5.6).

You could equally well have first gone the distance of y to reach the point Y, turn right and move a distance x. Then you would reach the same point P. That is our obvious everyday geometry. Not so in this strange micro world! You would reach quite different points in both the cases! (More

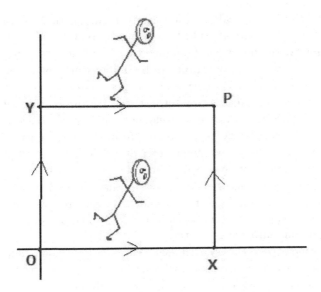

Fig. 5.6   Noncommutative geometry

precisely, the area of the rectangle first covered, OXPY will not be the same as the area you next cover, OYPX, something which is equally crazy.) This example is an over simplication of course, but the point is, in this example we are able to move along one straight line and then another. The units here are lengths. In the new thinking however, the units are areas, not so much lengths. We call this the quantum of area.

What all this means is we cannot go down to lower and lower scales arbitrarily. As we approach a minimum length we return to the larger universe! [57].

The interesting thing about Quantum superstring theory is the natural emergence of the spin 2 graviton As Witten a doyen of string theory puts it, the theory "predicts" gravitation. There appears a possibility for the elusive unified description.

Meanwhile an idea called supersymmetry or SUSY developed in parallel. This theory requires that each particle with integral spin has a counterpart with the same mass but having half integral spin. That is bosons have their supersymmetric counterparts in fermions. The result is that we try to describe fermions and bosons in a unified description. This leads to twice the number of particles as we have! SUSY is then broken like the symmetry breaking which lead to weak interactions or the magnetic North and South poles. So the counterparts would have a much greater mass, which would then account for the fact that these latter have not been observed. On the plus side in this theory gravitation can be unified with the other forces.

In fact this had lead to a theory called supergravity in which the spin 2 graviton takes on a spin 3/2 counterpart, the gravitino. Supergravity requires eleven spacetime dimensions, one more than superstring theory [58]. Unfortunately supergravity began to fade from the mid eighties because it could not explain other phenomena easily on reduction to the four physical spacetime dimensions from eleven. Quantum super string theory was in comparison more satisfactory. We could say that the earlier bosonic string theory worked in a spacetime that was bosonic, there being no place for spin. Quantum superstring works in a fermionic spacetime where we have the new non commutative geometry.

Quantum superstring theory displaced eleven dimensional supergravity. But even here there were many awkward questions. It turned out that there were many different ways in which we could go down from the original ten dimensions to our physical universe of four dimensions. After all, we cannot go around convincing one and all that our world is actually ten dimensional! However ingenious our explanation for the fact that we cannot perceive six of the ten dimensions, the explanation itself needs to be unique, at the very least.

After all we need a unique theory. And then why ten dimensions, while supersymmetry allows eleven dimensions? Another not very convincing factor was the fact that particles were being represented as one dimensional strings. Surely we could have two dimensional surfaces or membranes or any dimensional entities for that matter. This generalization resembles the earlier attempt of Dirac's, representing particles as bubbles or membranes. The theory also begins to encounter exotic types of spacetime, very different from what we can usually imagine. This is the realm of a topic called Topology.

Over the past few years, a variant of Quantum superstring theory has emerged called M Theory. This theory also uses supersymmetry but explains why the postulated super particles do not have the same mass as the known particles. Further these new masses must be much too heavy to be detected by current accelerators or collidors. The advantage of supersymmetry (SUSY) is that a framework is now available for the unification of all the interactions including gravitation. SUSY leaves the laws of physics the same for all observers, which is the case in general relativity (gravitation) also. Under SUSY there can be a maximum of eleven dimensions, the extra dimensions being curled up by the Kaluza-Klein trick.

Several mindboggling features emerge from all this. For instance, what would be energy in one universe would be electrical charge in another and so on.

Further the eleventh and extra dimension of the M-Theory could be shrunk, so that there would be two ten dimensional universes connected by the eleven dimensional spacetime. Now particles and strings would exist in the parallel universes which can interact through gravitation. The interesting aspect of the above scenario is that it is possible to conceive of all the four interactions converging at an energy far less than the Planck energy which is nearly a billion trillion times that of the proton. Infact the Planck energy is so high that it is beyond foreseeable experiments. Thus this would bring the eleven dimensional M-Theory closer to experiment.

So M-Theory is the new avatar of Quantum superstring theory. Nevertheless it is still far from being the last word. There are still any number of routes for compressing ten dimensions to our four dimensions. There is still no contact with experiment. It also appears that these theories suggest that the universe will not just expand, as it is now, but would actually blow up. This is because, the accelerated expansion of the universe is governed by what is called the cosmological constant, first introduced, though erroneously by Einstein, a century ago. This constant needs to be small. But M-Theory and variants give such a large cosmological constant, that a cosmic blow out is the only option.

An alternative approach was developed in the mid eighties by the Indian American physicist Abhay Ashtekar and other American physicists, Ted Jacobson, Lee Smolin, the Italian-French Carlo Rovelli and others. This has come to be known as Loop Quantum Gravity (LQG). In this approach too, spacetime is no longer smooth. Rather it is like a net. However there is an ingenious use of Quantum theory with concepts of General Relativity. Let us see what these are.

To put it roughly, the geometry of spacetime is now an evolving dynamical quantity, which can be obtained from suitable equations. Any set of coordinates we choose can be used to describe spacetime phenomena. Surprisingly the above considerations lead to the conclusion that space is not continuous but rather, discrete.

So in Loop Quantum Gravity, a volume in space or the surface of this volume cannot be shrunk to as small a size as we please - there are fundamental minimum units. The minimum unit of length is the Planck length, a billion, trillion trillionth of a centimeter! The minimum area turns out to be fundamental, you can think of it as a square with this length and breadth.

A volume is depicted by a dot while the enclosing flat surfaces are depicted by lines sticking out of the dot. You could think of it as the beads and wires changeable structures of the famous Indian toy. So any volume would be, graphically a network of these dots and lines (Cf.Fig.5.7), (Cf.Fig.5.8).

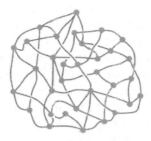

Fig. 5.7   Geometry as beads and wires

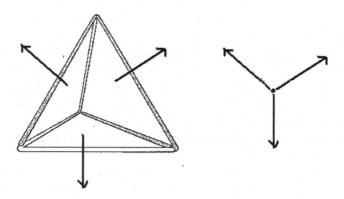

Fig. 5.8   A closer look at a "bead"

In fact the important idea is that this network of dots and lines *is* space rather than being something drawn in space. This is a bit like what we noted earlier – the actors are (or constitute) the stage, rather than being external objects on a pre given stage [59]. Every Quantum state corresponds to one of the possible networks formed by points or dots and lines. The dots and lines (or wires and beads) would also represent respectively particles and fields. Motion is now a result of discrete changes in this "wires and beads" toy. As time passes, the toy changes its shape in discrete steps. But this change is not predictable – and also there is nobody manipulating the toy. All this happens spontaneously.

Though Loop Quantum Gravity has made some progress over the years, as Lee Smolin, one of the founders puts it, "Many open questions remain to be answered in Loop Quantum Gravity. Some are technical matters that need to be clarified. We would also like to understand how, if at all, special relativity must be modified at extremely high energies. So far our speculations on this topic are not solidly linked to Loop Quantum Gravity calculations. In addition, we would like to know that classical general relativity is a good approximate description of the theory for distances much larger than the Planck length, in all circumstances. (At present we know only that the approximation is good for certain states that describe rather weak gravitational waves propagating on an otherwise flat spacetime.) Finally, we would like to understand whether or not Loop Quantum Gravity has anything to say about unification: Are the different forces, including gravity, all aspects of a single, fundamental force? String theory is based on a particular idea about unification, but we also have ideas for achieving unification with Loop Quantum Gravity."

5.5 A Critique

Reductionism has been at the heart of twentieth century theoretical physics. Beginning with the atomism of ancient Indian and Greek thinkers, it was reborn in the nineteenth century. This spirit is very much evident in Einstein's concept of locality in which an arbitrarily small part of the universe can be studied without reference to other parts of it. Indeed it is this philosophy of reductionism which has propelled the most recent studies such as string theory or other Quantum gravity approaches. However even after decades these have not revealed the final answers, we had hoped for. So a cascading chorus of dissenting voices is being heard in the corridors of science, questioning not only the approach, but even the very methodology of theoretical physics or more correctly the way it has been practised over the past few decades. Lee Smolin's very readable critique is to be found in his recent book "The Trouble With Physics" [60]. Smolin identifies some of the fundamental problems which we have to confront to make progress. This is tantamount to gripping the bull by the horns.

The first problem is the well known problem which we have mentioned namely to obtain a unified description of general relativity and Quantum theory. The second problem according to Smolin is the satisfactory resolution of the foundations of Quantum mechanics, "... either by making sense of the theory as it stands or by inventing a new theory that does make sense." As we have seen Quantum mechanics has continued to carry the baggage of an agreed upon interpretation. The third problem is to unify all the particles and fields in a single theory which brings them out as manifestations of a single fundamental entity. Indeed this is the paradigm of unification that we stressed. The fourth problem would be to, "explain how the values of the free constants in the standard model of particle physics are chosen in nature". We have already touched upon the fact that the standard model has to assign some eighteen different values to various constants, though it is not clear why these constants should have the values that we give, except of course the justification that this brings the theory into agreement with observation.

Finally we have to confront the identity and nature of dark matter and dark energy. As we saw the exact identity of dark matter has eluded us for many decades. The only justification for its existence was that it was necessary to explain observations (of galactic rotations). Dark energy is a recent entrant. Once again it explains observations – the acceleration of the universe – though its exact identity is unclear.

The fact is that these questions are much too big to be ignored. In fact they seem to be screaming out that a fundamental revision of our theories of space, time, matter and energy are called for. So far attempts have largely been tinkerings, a bit like the many epicycles and epi epicycles of the Greek model, merely to bring the model into line with observation.

In recent years, string theory and its descendent theories have been singled out for the type of bitter criticism which is unfamiliar in science. As we will see in a moment, the theorists have themselves to blame for over hyping their work, which otherwise is mathematically very rich. It is interesting to see what two brilliant physicists of the previous generation had to say on this topic. Professor Abdus Salam was a great enthusiast. In his words, [61] "This is something which has happened since 1985. From that time we have begun to believe not in particles but in strings... So these strings are the fundamental entities nowadays. We believe in them... This gives us a theory which is free of infinities, free of inconsistencies and its a beautiful theory..."

On the other hand the celebrated Richard Feynman was very pessimistic about string theories and believed that the entire endeavor would come to nothing. As he put it, "I don't like that they're not calculating anything. I don't like that they don't check their ideas. I don't like that for anything that disagrees with an experiment, they cook up an explanation – a fix up to say 'Well, it still might be true.' For example, the theory requires ten dimensions. Yes, that's possible mathematically, but why not seven? When they write their equation, the equation should decide how many of these things get wrapped up, not the desire to agree with experiment. In other words, there's no reason whatsoever in super-string theory that it isn't eight of the ten dimensions that get wrapped up and that the result is only two dimensions, which would be completely in disagreement with experience. So the fact that it might disagree with experience is very tenuous, it doesn't produce anything it has to be excused most of the time. It doesn't look right."

Against this backdrop, the first salvo was fired by Nobel Laureate R.B. Laughlin. "A Different Universe", his recent book [62], would come as a shock because he debunks reductionism in favour of what is these days called emergence. That is his central theme. The fundamental laws of nature emerge through collective self organization and do not require knowledge of their component parts, that is microscopic rules, in order that we comprehend or exploit them. The distinction between fundamental laws and laws descending from them is a myth. In his words, "... I must openly discuss some shocking ideas: the vacuum of space-time is 'matter', the possibility that relativity is not fundamental..." He argues that all fundamental constants require an environmental context to make sense. This is contrary to the reductionist view that basic bricks build up structures.

Laughlin takes pain to bring to our notice that there is now a paradigm shift from the older reductionist view to a view of emergence. For him Von Klitzing's beautiful experiment bringing out the Quantum Hall effect is symbolic of the new ethos. The Quantum of Hall resistance is a combination of fundamental constants viz., the indivisible quantum of electric charge e, the Planck constant h and the speed of light c. This means that these supposedly basic building blocks of the universe can be measured with breathtaking accuracy, without dealing with the building blocks themselves. Though, from one point of view, this resembles the fact that bulk properties emerge from underlying and more fundamental microscopic properties, he argues that this latter effect reveals that supposedly indivisible quanta like the electric charge e can be broken into pieces through a process of self organization. That is, the supposedly fundamental things are not necessarily fundamental. Furthermore, for example, in superconductivity, many of the so called minor details are actually inessential - the exactness of the Meissner and Josephson effects does not require the rest of the finer detail to be true.

Admittedly there are a number of grey areas in modern theoretical physics which are generally glossed over. For instance Dirac's Hole Theory of anti matter which we encountered earlier. However in silicon, there are many electrons locked up in the chemical bonds and it is possible to pull an electron out of a chemical bond. This makes a hole which is mobile and acts in every way like an extra electron with opposite charge added to the silicon. This idea however requires the analogue of a solid's bond length. In particle physics such a length conflicts fundamentally with the principle of relativity as we have seen, unless, as we have argued, it breaks down at the Compton scale. On the contrary, Laughlin laments, "... instead, physicists have developed clever semantic techniques for papering it over... Thus instead of Holes one speaks of anti particles. Instead of bond length one speaks of an abstraction called the ultra violet cutoff, a tiny length scale introduced into the problem to regulate - which is to say, to cause it to make sense. Below this scale one simply aborts one's calculations... Much of Quantum Electrodynamics, the mathematical description of how light communicates with the ocean of electrons... boils down to demonstrating the unmeasurableness of the ultra violet cutoff... The potential of overcoming the ultra violet problem is also the deeper reason for the allure of String Theory, a microscopic model for the vacuum that has failed to account for any measured thing... The properties of empty space relevant to our lives show all the signs of being emergent phenomena characteristic of a phase of matter. They are simple, exact, model insensitive, and universal. This is what insensitivity to the ultra violet cutoff means physically."

Moreover quantized sound waves or phonons have an exact parallel with photons - in fact their quantum properties are identical to those of light. However sound is a collective motion of elastic matter, while in our understanding, light is not. This means that quantization of sound may be deduced from the underlying laws of Quantum mechanics obeyed by the atoms, whereas in the case of light this is postulated. This is a logical loose end and ultimately we bring in the gauge effect to cover this. But unfortunately, "there is also a fundamental incompatibility of the gauge effect with the principle of relativity, which one must sweep under the rug by manipulating the cutoff." Laughlin complains that in spite of the evidence against reductionism, sub nuclear experiments are generally described in reductionist terms. In any case, we will return to these ideas in the next Chapter.

Turning to general relativity, Laughlin points out that it is a speculative post Newtonian Theory of Gravity, an invention of the mind, "it is just controversial and largely beyond the reach of experiment", unlike special relativity which was a discovery of the behavior of nature. He then points out the contradiction between special and general relativity - in the former Einstein did away with the concept of the ether. But this reenters the latter theory in the form of the fabric of space. Touching upon the skeletons in the closet of general relativity, Laughlin discusses the embarrassment that is caused by a non zero cosmological constant, that is, that the universe is accelerating, instead of decelarating under its own weight.

He concludes that if Einstein were alive today, he would be horrified at this state of affairs and would conclude that his beloved principle of relativity was not fundamental at all but emergent.

Laughlin takes a critical look at renormalizability, a pillar of modern theoretical physics, and cosmology. Indeed, we have already commented on this. "If renormalizability of the vacuum is caused by proximity to phase transitions, then the search for an ultimate theory would be doomed on two counts: It would not predict anything even if you found it, and it could not be falsified..."

Indeed Dirac, one of the early pioneers of the renormalization programme expressed his pessimism about this approach and went on to say that its success may be a fluke. Renormalization the all important sophisticated technique for bypassing annoying infinities in Quantum Field Theory has been a not very satisfying feature. Quite a few theorists have been uncomfortable with it. As Hawking puts it mildly, "...This involves cancelling the infinities by introducing other infinities. Although this technique is rather dubious mathematically, it does seem to work in practice, and has been used with these theories to make predictions that agree with observations to an extraordinary degree of accuracy. Renormalization, however, does have a serious drawback from the point of view of trying to find a complete theory, because it means that the actual values of the masses and the strengths of the forces cannot be predicted from the theory, but have to be chosen to fit the observations." [63]

Laughlin's critique of string inspired cosmology is equally stinging. " The political nature of cosmological theories explains how they could so easily amalgamate String Theory, a body of mathematics with which they actually have very little in common... (String Theory) has no practical utility however, other than to sustain the myth of the ultimate theory. There is no experimental evidence for the existence of Strings in nature... String Theory is, in fact a textbook case of ... a beautiful set of ideas that will always remain just barely out of reach. Far from a wonderful technological hope for a greater tomorrow, it is instead a tragic consequence of an absolute belief system in which emergence plays no role..."

Laughlin has captured the mood of pessimism that prevails in the minds of several high energy physicists. He goes on to cite the famous joke that the hallowed Physical Review is now so voluminous that stacking up successive issues would generate a surface travelling faster than the speed of light, although without violating relativity, because the Physical Review contains no information anyway.

However my own work during the last decade plus has borne out the spirit of these ideas, that the iron clad law of physics is more thermodynamic and probabilistic in nature, that the velocity of light or the gravitational constant can be deduced from such considerations rather than be taken as fundamental inputs; how, it is possible to have schemes that bypass the awkward questions raised in the book, without brushing them away below the carpet, by considering an a priori Quantum vacuum in which fluctuations take place. We will examine this a little later.

Returning now to String Theory there is no doubt that it has straddled the past two decades and more as the only contender for the Theory of Everything. Some years ago Nobel Laureate Sheldon Glashow described it, sarcastically, as the only game in town. In recent years though the theory is not only being debunked, it is facing a lot of flak, particularly in the worldwide media. Laughlin may be faulted on the grounds that he is not a string theorist or a particle physicist. The decisive tilt has come from Nobel Laureate David Gross, very much an insider, who as it were, spilt the beans at the 23rd Solvay Conference in Physics held in Brussels, Belgium, in late 2005. He stated "We don't know what we are talking about." He then went on to say, "Many of us believed that string theory was a very dramatic break with out previous notions of quantum theory. But now we learn that string theory, well, is not that much of a break." He added that physics is in "a period of utter confusion."

At this meeting Gross compared the state of physics today to that during the first Solvay Conference in 1911 "They were missing something absolutely fundamental," he said. "We are missing perhaps something as profound as they were back then."

Let me try to give you an idea of the popular mood in recent times. The Time Magazine, August 14, 2006 issue notes:

"By now, just about everyone has heard of string theory. Even those who don't really understand it – which is to say, just about everyone – know that it's the hottest thing in theoretical physics. Any university that doesn't have at least one string theorist on the payroll is considered a scientific backwater. The public, meanwhile, has been regaled for years with magazine articles

"But despite its extraordinary popularity among some of the smartest people on the planet, string theory hasn't been embraced by everyone-and now, nearly 30 years after it made its initial splash, some of the doubters are becoming more vocal. Skeptical bloggers have become increasingly critical of the theory, and next month two books will be hitting the shelves to make the point in greater detail. Not Even Wrong, by Columbia University mathematician Peter Woit, and The Trouble With Physics, by Lee Smolin at the Perimeter Institute for Theoretical Physics in Waterloo, Ont., both argue that string theory (or superstring theory, as it is also known) is largely a fad propped up by practitioners who tend to be arrogantly dismissive of anyone who dare suggest that the emperor has no clothes

"Bizarre as it seemed, this scheme appeared on first blush to explain why particles have the characteristics they do. As a side benefit, it also included a quantum version of gravity and thus of relativity. Just as important, nobody had a better idea. So lots of physicists, including Woit and Smolin, began working on it.

"Since then, however, superstrings have proved a lot more complex than anyone expected. The mathematics is excruciatingly tough, and when problems arise, the solutions often introduce yet another layer of complexity. Indeed, one of the theory's proponents calls the latest of many string-theory refinements "a Rube Goldberg contraption." Complexity isn't necessarily the kiss of death in physics, but in this case the new, improved theory posits a nearly infinite number of different possible universes, with no way of showing that ours is more likely than any of the others.

"That lack of specificity hasn't slowed down the string folks. Maybe, they've argued, there really are an infinite number of universes-an idea that's currently in vogue among some astronomers as well-and some version of the theory describes each of them. That means any prediction, however outlandish, has a chance of being valid for at least one universe, and no prediction, however sensible, might be valid for all of them.

"That sort of reasoning drives critics up the wall. It was bad enough, they say, when string theorists treated nonbelievers as though they were a little slow-witted. Now, it seems, at least some superstring advocates are ready to abandon the essential definition of science itself on the basis that string theory is too important to be hampered by old-fashioned notions of experimental proof

"And it is that absence of proof that is perhaps most damning."

"It's fine to propose speculative ideas," says Woit, "but if they can't be tested they're not science." To borrow the withering dismissal coined by the great physicist Wolfgang Pauli, they don't even rise to the level of being wrong. That, says Sean Carroll of the University of Chicago, who has worked on strings, is unfortunate. "I wish string theorists would take the goal of connecting to experiment more seriously," he says.

Lee Smolin did early work in string theory but drifted away to Loop Quantum Gravity. Standard particle physics has, as we saw several shortcomings. Another shortcoming has to do with cosmology. According to standard theory, the universe should be accelerating so fast that it would be nothing short of a blow out, and we should not be existing. This is well known as the cosmological constant problem [64]. It arises because all the elementary particles have a lowest possible energy level. However as we saw, Quantum theory prohibits such precise values. There is a fluctuation about these values, called vacuum energy. There are so many elementary particles in the universe, that the sum total of these vacuum energies would be incalculably enormous. It is this huge energy that would rip the universe apart. As Smolin notes in his book, "The cosmological constant posed a problem for all of physics, but the situation appeared a bit better for string theory. String theory could not explain why the cosmological constant was zero, but at least i t explained why it was not a positive number (that is why the universe would not accelerate). One of the few things we could conclude from the string theories then known was that the cosmological constant could only be zero or negative..."

"You can imagine the surprise, then, in 1998 when the observations of supernovas began to show that the expansion of the universe was accelerating, meaning that the cosmological constant had to be a positive number. This was a genuine crisis because there appeared to be a clear disagreement between observation and a prediction of string theory...

"Edward Witten is not someone given to pessimism, yet he flatly declared in 2001, that 'I don't know any clear cut way to get (a universe with a positive cosmological constant) from String Theory or M Theory'".

The situation is actually murkier. There could be a concocted theory with a positive cosmological constant, except that there are trillions and trillions of such theories or solutions. To be precise, the number would be one followed by five hundred zeroes! Enough zeroes to fill a page. Far too many to be taken seriously. Yet, scientists don't give up easily as we saw earlier too.

In recent years it is being argued that there is nothing wrong in these many possibilities. In fact all these could constitute a huge collection of different universes, or a landscape of universes, each with its own distinct characteristic defined by a definite possibility. In other words the various universes would have very different laws and features. It so happens, the argument goes that we are inhabiting one of this multitude of these universes in which the laws are what they are. This is called anthropic reasoning, and was initiated some decades ago in a different context amongst others, by the celebrated Stephen Hawking [65]. For many such an argument is equivalent to the statement, don't ask why things are what and as they are. It just so happens that's how things are. In any case, you wouldn't be there to ask these questions if things were otherwise! Put this way, the argument trivializes the very spirit of science.

David Gross, who got the Nobel Prize for his work on the standard model and later contributed significantly to string theory himself admitted: "We see this kind of thing happen over and over again as a reaction to difficult problems... Come up with a grand principle that explains why you're unable to solve the problem." Indeed this illustrates a comment by Richard Feynman, who himself was not in favor of string theory as already noted, "String theorists make excuses, not predictions."

According to the August 27, 2006 issue of the Scientific American,

"With a tweak to the algorithms and a different database, the Website could probably be made to spit out what appear to be abstracts about superstring theory: 'Frobenius transformation, mirror map and instanton numbers' or 'Fractional two-branes, toric orbifolds and the quantum McKay correspondence.' Those are actually titles of papers recently posted to arXiv.org repository of preprints in theoretical physics, and they may well be of scientific worth-if, that is, superstring theory really is a science. Two new books suggest otherwise: that the frenzy of research into strings and branes and curled-up dimensions is a case of surface without depth, a solipsistic shuffling of symbols as relevant to understanding the universe as randomly generated dadaist prose.

"In this grim assessment, string theory-an attempt to weave together general relativity and quantum mechanics-is not just untested but untestable, incapable of ever making predictions that can be experimentally checked. With no means to verify its truth, superstring theory, in the words of Burton Richter, director emeritus of the Stanford Linear Accelerator Center, may turn out to be "a kind of metaphysical wonderland." Yet it is being pursued as vigorously as ever, its critics complain, treated as the only game in town.

'String theory now has such a dominant position in the academy that it is practically career suicide for young theoretical physicists not to join the field,' writes Lee Smolin, a physicist at the Perimeter Institute for Theoretical Physics, in The Trouble with Physics: The Rise of String Theory, the Fall of a Science, and What Comes Next. 'Some young string theorists have told me that they feel constrained to work on string theory whether or not they believe in it, because it is perceived as the ticket to a professorship at a university.'

"Neither of these books can be dismissed as a diatribe. Both Smolin and Woit acknowledge that some important mathematics has come from contemplating superstrings. But with no proper theory in sight, they assert, it is time to move on. 'The one thing everyone who cares about fundamental physics seems to agree on is that new ideas are needed,' Smolin writes. 'We are missing something big.'

"The story of how a backwater of theoretical physics became not just the rage but the establishment has all the booms and busts of an Old West mining town. Unable to fit the four forces of nature under the same roof, a few theorists in the 1970s began adding extra rooms-the seven dimensions of additional closet space that unification seemed to demand. With some mathematical sleight of hand, these unseen dimensions could be curled up ("compactified") and hidden inside the cracks of the theory, but there were an infinite number of ways to do this. One of the arrangements might describe this universe, but which?

"The despair turned to excitement when the possibilities were reduced to five and to exhilaration when, in the mid-1990s, the five were funneled into something called M Theory, which promised to be the one true way. There were even hopes of experimental verification

"That was six years ago, and to hear Smolin and Woit tell it, the field is back to square one: recent research suggests that there are, in fact, some

10^{500} perfectly good M theories, each describing a different physics. The theory of everything, as Smolin puts it, has become a theory of anything. "Faced with this free-for-all, some string theorists have concluded that there is no unique theory, that the universe is not elegant but accidental. If so, trying to explain the value of the cosmological constant would make as much sense as seeking a deep mathematical reason for why stop signs are octagonal or why there are 33 human vertebrae"

An article in the Financial Times (London) in June 2006 by Physicist Robert Mathews noted:

"They call their leader The Pope, insist theirs is the only path to enlightenment and attract a steady stream of young acolytes to their cause. A crackpot religious cult? No, something far scarier: a scientific community that has completely lost touch with reality and is robbing us of some of our most brilliant minds.

"Yet if you listened to its cheerleaders-or read one of their best-selling books or watched their television mini-series-you, too, might fall under their spell. You, too, might come to believe they really are close to revealing the ultimate universal truths, in the form of a set of equations describing the cosmos and everything in it. Or, as they modestly put it, a 'theory of everything'.

"This is not a truth universally acknowledged. For years there has been concern within the rest of the scientific community that the quest for the theory of everything is an exercise in self-delusion. This is based on the simple fact that, in spite of decades of effort, the quest has failed to produce a single testable prediction, let alone one that has been confirmed.

"For many scientists, that makes the whole enterprise worse than a theory that proves to be wrong. It puts it in the worst category of scientific theories, identified by the Nobel Prize-winning physicist Wolfgang Pauli: it is not even wrong. By failing to make any predictions, it is impossible to tell if it is a turkey, let alone a triumph.

"It is this loss of contact with reality that has prompted so much concern among scientists-at least, those who are not intimidated by all the talk of multidimensional superstrings and Calabi-Yau manifolds that goes with the territory. But now one of them has decided the outside world should be told about this scientific charade. As a mathematician at Columbia University, Peter Woit has followed the quest for the theory of everything for more than 20 years. In his new book 'Not Even Wrong' [66] he charts how a once-promising approach to the deepest mysteries in science has mutated into something worryingly close to a religious cult."

Peter Woit notes, "No other specific speculative idea about particle physics that has not yet been connected to anything in the real world has ever received anywhere near this amount of attention." Elsewhere he quotes Thomas Harris in Hannibal, "There are repeated efforts with the symbols of string theory. The few mathematicians who could follow him might say his equations begin brilliantly and then decline, doomed by wishful thinking." Indeed to quote Woit again, "No matter how things turn out, the story of super string is an episode with no real parallel in the history of modern physical science. More than twenty years of intensive research by thousands of the best scientists in the world producing tens of thousands of scientific papers has not lead to a single testable experimental prediction of the theory... The question asks whether the theory is a science at all."

A review in the December 2006 issue of Physics Today notes: "Noted theoretical physicist Sheldon Glashow has famously likened string theory to medieval theology because he believes both are speculations that cannot be tested. Yet if readers believe Lee Smolin and Peter Woit, they might conclude that the more apt comparison is to the Great disappointment of 1844, when followers of the Baptist preacher William Miller gave up all their worldly possessions and waited for the Second Coming. The empirical inadequacy of that prediction led to apostasy and schisms among thousands of Miller's followers. At least one of the branches claimed that the event had in fact occurred, but in a heavenly landscape linked to the world of experience through only the weak but all-pervasive spiritual interaction. Yet irritating differences exist between Miller's followers and the "disappointed" of the 1984 coming of the theory of everything; a majority of the latter seem to have preserved their faith and gained worldly fortune in the form of funding, jobs, and luxurious conferences at exotic locales."

In all this confusion, we should not forget two important points. The first is that string theory still remains a mathematically beautiful self consistent system of thought. Perhaps the flak that string theory is receiving is more due to reasons in the domain of the sociology of science. To elaborate, string theory has been touted as a theory, which, in the strict sense it is not. If it had been promoted as a hypothesis, one of a few possible, perhaps, there would have been much less criticism.

Indeed, some years ago Nobel Laureate 't Hooft had noted "Actually, I would not even be prepared to call string theory a "theory" but rather a 'model' or not even that: just a hunch. After all, a theory should come together with instructions on how to deal with it to identify the things one wishes to describe, in our case the elementary particles, and one should, at least in principle, be able to formulate the rules for calculating the properties of these particles, and how to make new predictions for them"

As Lee Smolin puts it, "... I can think of no mainstream string theorist who has proposed an original idea about the foundations of Quantum Theory or the nature of time." What he means is, that there is a lot of technical mathematics, but hardly any conceptual advance in all this.

Moreover in the process string theory adopted strong arm facist type tactics including marketing through the media, while at the same time making not too covert attempts to suppress other ideas. The Late Yuval Ne'eman, one of the earliest proponents of the quark theory of elementary particles, once told me that you cannot get any article published in the top journals, or even obtain a decent research fellowship, if you did not work on string theory. He went on to say that eventually this unethical strangle hold would come loose. The backlash was therefore inevitable.

The second point is even more important and is expressed in David Gross's statement that perhaps we are missing something very profound. This throws up a great challenge and makes for very exciting times.

Science has been described as a quest for the how and why of nature. Over the centuries it has been guided by some principles which have crystallized into a methodology. Thus observation leads to the framing of hypotheses. It is expected that the hypotheses would have maximum simplicity and maximal economy. This means that, apart from being simple, the maximum number of observations are explained by a minimum number of hypotheses. Further tests would then confirm or disprove the hypotheses, if the hypotheses are found to be consistent with experiment. The richness of a hypothesis is judged by the predictions it can make. These predictions must be either provable or disprovable as stressed by Sir Karl Popper. As regards the poorer cousin of string theory, LQG, its proponents do indeed concede that there is still some way to go. Some critics are even less enthusiastic about it, compared to string theory. They point out that string theory has atleast thrown up rich mathematical structures, unlike LQG. In any case there is now a hope that alternative approaches will also get a hearing. That is very essential, a point made by the celebrated Lord Kelvin more than a century ago. To understand why, just imagine what would have happened if dinosaurs were the only species on the earth sixty five million years ago. Their extinction would have left the earth a lifeless planet!

As far as fundamental physics is concerned, starting from the early days of Indian and later Greek atomism, through the Atomic Theory of the

19th century, and subsequently the developments in the 20th century, and the early part of the 21st century, the route followed has been one of a descending-in-size cascade. We have been propelled by the belief that the universe could be understood by a study of its ultimate subconstituents. This spirit as noted is very much evident in Einstein's concepts of locality in which an arbitrarily small part of the universe can be studied without reference to other parts of it. A few decades later Wheeler observed that our studies of the inaccessible Planck scale of $10^{-33}cms$ were really like an understanding of bulk properties of matter by studying the subconstituent molecules.

Indeed it is this philosophy of reductionism which has propelled the most recent studies such as string theory or other Quantum gravity approaches. Decades of labour has gone into these endeavours and the research output has been enormous.

Nevertheless we seem to have reached an impasse of a type that is all too familiar from the past. There are minor discrepancies or corrections, which nevertheless would point to, not just an incremental change of our concepts, but rather to a paradigm shift itself of the type we witnessed with Kepler's ellipses or Einstein's Relativity or Quantum Theory. Any theory can go only so far as its inherent limitations or constraints permit. At that stage, as Thomas Kuhn notes, there would be a revolution, an overturn of concepts and the old way of looking at things. It is no longer an incremental improvement. We will see this in a little more detail in the next Chapter.

5.6   Smashing the Atoms

As we noted we can get information about the most fundamental foundations of the universe in the form of its most elementary constituents, the elementary particles and the different types of forces which exist amongst the particles, by studying the collision of particles. We could observe this drama of elementary particles using accelerators or collidors. As the name suggests, charged particles are accelerated using magnetic fields. As these particles travel very fast, the accelerators would need to be very large - unrealistically large in theory. In practise what we do is, we allow these particles to go round and round in a circular tube, and accelerate them periodically. This way we can get really fast moving charged particles in a realistically confined space. Moreover we can get these particles to be traveling in opposite directions so that they collide, yielding vital information about the forces that operates at this level. The earlier collidor at CERN, the European Centre for Nuclear Research in Geneva was the SPP/S Collidor. This accelerated particles to nine hundred giga electron volts, half a million electron volts being the energy content of the electron. This is hundreds of times the energy carried by a proton traveling with nearly the speed of light. There is also the tevatron at Fermi Lab in Illinois. This can accelerate particles to more than double the energy of the older CERN machine. Similarly a few other accelerators have been operating like the Tristan machine in Japan or the Stanford Linear Accelerator Collidor (SLAC). These latter accelerators do not produce particles that are energetic enough to yield significant results. On the whole the American machines have lagged behind their European counterpart, in particular the electron positron collidor at CERN, with a circumference of twenty seven kilometers.

There was a bold attempt to use the new technology of superconductivity to build a superconductivity collidor in the United States at a cost of a few billion dollars. This would have had a circular tunnel with a circumference of nearly a hundred kilometers and it was envisaged that it would produce particles with energies like trillions of electron volts. But the proposal was not accepted.

You may wonder, what is the difference between accelerators and collidors. Previously we would accelerate charged particles and allow them to strike a stationary target–for example the nucleus of an atom in the target. This is reminiscent of Rutherford's gold foil experiment, but has its limitations because the stationary target too would carry away some of the energy. With collidors this disadvantage is eliminated because we have two particles moving in opposite directions that smash into each other, and each particle can be observed. However, there is a disadvantage too. And that is that the number of particles in the beams available for collisions is much less than the number of particles in a stationary target.

Even in collidors there are two kinds: Those in which particles like electrons and positrons collide and those in which the much heavier protons collide. In proton-proton collisions there is a lot of unnecessary information that is generated. This is because the protons are made up of three quarks each. We would ideally like to study the quark-quark collisions but because they are hidden in the protons, there is no direct quark-quark collision.

On the other hand electrons are not composites–they are truly elementary and so we can clearly observe electron-positron collisions without any complications. The disadvantage however is that electrons and positrons being so much lighter than protons can get deflected by magnetic fields more easily–in the process they loose a lot of energy and this reduces the acceleration. For electron-positron collidors therefore we need to have many more of weaker magnets producing weaker magnetic fields. This in turn will increase the length of the machine.

All existing accelerators and collidors however fail in comparison to the recently, nearly inaugurated Large Hadron Collidor at CERN, in September 2008.

The Large Hadron Collidor on the outskirts of Geneva, bordering Switzerland and France is the largest and most expensive scientific experiment to-date. The twenty seven kilometer tunnel is buried fifty to a hundred and seventy five meters below the ground. What will happen here is that protons will be made to collide with protons, each of the beams rushing with an energy of some seven trillion electron volts, once the maximum performance is achieved. This is far more than the output of any of the existing collidors. Scientists have huge expectations and hopes hanging on the LHC.

The standard model of particle physics reached a peak in the early seventies, particularly with the unification of the electromagnetic and weak interactions, for which as we noted Steven Weinberg and Sheldon Glashow and Abdus Salam were given the Nobel Prize. But this does not mean that the standard model has answered all the questions. As we saw there are still many gaps to be filled. Moreover physicists are now convinced that we have to look for a theory that goes beyond the standard model. This has been necessitated by the fact that in the standard model the neutrino has no mass. However in the late nineties the SuperKamiokande experiment in Japan conclusively proved that the neutrino has a small mass. So, clearly this is not the whole story. The LHC has to throw up the Higgs particle which is one of the missing pieces of the jigsaw puzzle. It is so important because it is the agent that in the theory provides the mass to all the particles. Then there is a question of why the elementary particles have the masses which they do have? In fact today we need to explain eighteen parameters as the physicist would say.

The particle beams in the LHC ring travel at nearly the speed of the light, in a vacuum, guided by superconducting magnets which are cooled by a huge cryogenic system. Because of the superconducting feature the electricity flows without any resistance. These interacting particles are stored for a few hours, something which consumes a huge amount of energy. Can they throw up the answers to questions which have dogged physicists for a few decades? Can the Higgs particle be found at last?

It is of course to be expected that whatever answers the LHC may or may not yield, it would throw up many new particles. Who knows these may actually complicate the situation by pointing at new mysteries. In any case the LHC after many delays was launched on the 10th September 2008. After a few days, on the 19th of September, due to an unfortunate accident, several of the superconducting magnets got damaged. The repairs would take time and cost money. Today scientists do not expect any new results from the LHC till 2010. In any case as an International publication put it, "Particle physics is the unbelievable pursuit of the unimaginable. To pinpoint the smallest fragments of the universe you have to build the biggest machine in the world. To recreate the first millionth of the second of creation, you have to focus energy on an awesome scale."

In any case the short-lived inauguration of the LHC itself was preceded by unease and even protests in some quarters. It was rumored that the big bang recreation in the machine would blow up the universe and unsuccessful attempts at obtaining court injunctions were made.

In a sense we are putting on a bold face. Our efforts do not go far enough. For testing even higher energies which are thousands of trillions of TeV, we may perhaps need to have an accelerator that is so large that it circles the earth. Professor Abdus Salam would say, "theoretically if you want Planck energy then you would need an accelerator ten light years long. So ten light years means that you will have dynasties of people who will go on various spacecrafts and one father would say to the son: My dear son I did this experiment up to a certain stage. Will you carry on from there? So that's what we are coming to."

Chapter 6

Law without law

"Traveller, there are no paths,
Paths are made by the walking"

– Antonio Machado.

6.1 The Perfect Universe

One Summer day in 1595 Johannes Kepler, then an ill paid astronomy instructor was taking a class in Graz. He was a poor teacher and did not have much of communication with his students. But as he was scribbling, suddenly the secret key of the universe fell into his lap. He realized why there were only six planets and no more nor no less. The answer had to do with what Plato preached in his famous school almost two thousand years earlier. There were five perfect solids or "polyhedra" [2]. These are the so called Platonic solids. They include the cube with six faces, the tetrahedron with four faces, the dodecahedron and so on. These were the only possible solids with equal faces (Cf.Fig.6.1). The cube for example has six equal faces. What better scaffolding would the creator have than these perfect solids, to be inserted in the five spaces between two adjacent spheres, each carrying for example a planet? (What is left unsaid in this argument is, that having run out of perfect solids, the Creator stopped further creation.)

Let us see how Kepler's idea works. Let us fit the orbit of the earth into a sphere. Then we fix a dodecahedron around the sphere and surround it with another sphere containing the orbit of Mars. We then put the sphere of Mars into a tetrahedron which again is surrounded by a sphere, allotted for Jupiter. Next we use a cube for Saturn. There were still the Planets

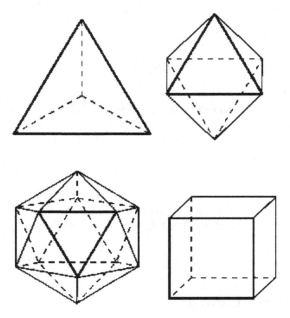

Fig. 6.1   Some perfect solids

Mercury and Venus to be accommodated. So let's place an icosahedron inside the earth's sphere and allot Venus to it. Finally inside the sphere of Venus we could fit in an octahedron for Mercury. This completed the list. Kepler who had a feverish imagination, immediately authored a book, "Mysterium Cosmographicum", incorporating these ideas. The year was 1596.

There could well be a parallel in today's science. From the beginning of modern science, the universe has been considered to be governed by rigid laws which therefore, in a sense, made the universe somehow determin- istic. The quest of science, as we saw in the previous Chapters, was a quest to find these laws and minimize their number. However, it would be more natural to expect that the underpinning for these laws would be random,unpredictable and spontaneous events rather than enforced before- hand phenomena. Otherwise, you could ask, why has nature preferred this rule, for example Newton's inverse square law, over that, for instance an inverse cubed law. There is a counter argument, though. You could invoke an "anthropic" argument and point out that if the laws had been different, so would the universe and then we would not be there to ask the question in the first place.

Returning to randomness, the question is, how can this be? Is there any mechanism leading from randomness to order? We have already seen an example. Water at room temperature is a liquid. The molecules are all jiggling about at random, buffeting each other, as noted by Perrin. If the water is cooled, at freezing point, this scenario of randomness suddenly gets transformed. We get ice with its distinctive crystallized form. This alternative but historical school of thought is in the spirit of Prigogine's, "Order out of chaos".

Prigogine notes [34] and we quote again, though a little more elaborately, "As we have already stated, we subscribe to the view that classical science has now reached its limit. One aspect of this transformation is the discovery of the limitations of classical concepts that imply that a knowledge of the world "as it is" was possible. The omniscient beings, Laplace's or Maxwell's demon, or Einstein's God, beings that play such an important role in scientific reasoning, embody the kinds of extrapolation physicists thought they were allowed to make. As randomness, complexity, and irreversibility enter into physics as objects of positive knowledge, we are moving away from this rather naive assumption of a direct connection between our description of the world and the world itself. Objectivity in theoretical physics takes on a more subtle meaning. ...Still there is only one type of change surviving in dynamics, one "process", and that is motion... It is interesting to compare dynamic change with the atomists' conception of change, which enjoyed considerable favor at the time Newton formulated his laws. Actually, it seems that not only Descartes, Gessendi, and d'Alembert, but even Newton himself believed that collisions between hard atoms were the ultimate, and perhaps the only, sources of changes of motion. Nevertheless, the dynamic and the atomic descriptions differ radically. Indeed, the continuous nature of the acceleration described by the dynamic equations is in sharp contrast with the discontinuous, instantaneous collisions between hard particles. Newton had already noticed that, in contradiction to dynamics, an irreversible loss of motion is involved in each hard collision. The only reversible collision – that is, the only one in agreement with the laws of dynamics – is the "elastic," momentum-conserving collision. But how can the complex property of "elasticity" be applied to atoms that are supposed to be the fundamental elements of nature?

"On the other hand, at a less technical level, the laws of dynamic motion seem to contradict the randomness generally attributed to collisions between atoms. The ancient philosophers had already pointed out that any natural process can be interpreted in many different ways in terms of the motion of and collisions between atoms."

If we follow this philosophy, then in the words of Wheeler, we seek ultimately a "Law without Law." Laws which physicists seek, are an apriori blue print within the constraints of which, the universe evolves. The point can be understood again in the words of Prigogine

"...This problem is a continuation of the famous controversy between Parmenides and Heraclitus. Parmenides insisted that there is nothing new, that everything was there and will be ever there. This statement is paradoxical because the situation changed before and after he wrote his famous poem. On the other hand, Heraclitus insisted on change. In a sense, after Newton's dynamics, it seemed that Parmenides was right, because Newton's theory is a deterministic theory and time is reversible. Therefore nothing new can appear. On the other hand, philosophers were divided. Many great philosophers shared the views of Parmenides. But since the nineteenth century, since Hegel, Bergson, Heidegger, philosophy took a different point of view. Time is our existential dimension. As you know, we have inherited from the nineteenth century two different world views. The world view of dynamics, mechanics and the world view of thermodynaics."

It may be mentioned that subsequent developments in Quantum theory, including Quantum Field Theory are in the spirit of the former world view. In this sense, they are extensions of Newtonian thinking, even though, as we saw, Quantum theory speaks of the unpredictable collapse of the wave function. The point is that the equations for these wave functions are deterministic! Einstein himself believed in this view of what may be called deterministic time - time that is also reversible. On the other hand Heraclitus's point of view was in the latter spirit. His famous dictum was, "You never step into the same river twice", a point of view which was anticipated by an earlier ancient Indian stream of thought [67].

This as we saw, has been the age old tussle between "being" and "becoming". "Being" can be likened to an unchanging but flowing river. "Becoming" on the other hand is in the spirit of Heraclitus – everytime you step into the river, you encounter different waters. Nevertheless, as we have seen, neither Quantum theory nor general relativity have had satisfactory closures, apart from the fact that their unified description has been elusive, even illusive. Indeed starting from the 1930s, a few serious physicists began to ask if there was something fundamentally incorrect about our concepts of space and time. Perhaps a radical departure was needed? It became clear, at least to some physicists that not just a cosmetic, but a rather profound and surgical revision of earlier concepts of physics was needed.

There are at least two concepts which we can revisit in this context. The first is, as we just now saw, not continuous space and time, but rather space and time which are more like a fisherman's net or to take another example, a sponge, with a lot of gaps and holes inside (Cf.Fig.6.2). The justification for this is Quantum theory itself. We cannot go down to points in a smooth spacetime without encountering meaningless infinite energies. But this is one side of the coin.

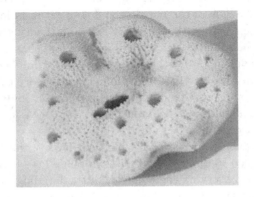

(a) Sponge

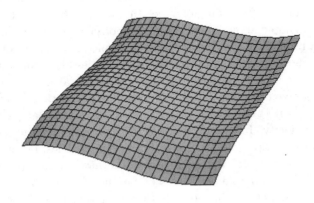

(b) Fisherman's net

Fig. 6.2   Sponge and fisherman's net

The other feature, which has been implicit in much of twentieth century thinking, is that of reductionism – to understand the large in terms of minute building blocks, going right down to the level of quarks, electrons and neutrinos. Given these building blocks, we can construct the edifice. That has been the dominant strain of the philosophy of fundamental physics. You would notice that the edifice itself, does not have any effect on the building blocks. Or to put it another way the environment has no role on the characteristics of the building blocks which can therefore be studied quite independently. Einstein firmly believed in two concepts – locality and reality. The former meant, as we saw, that we can pinch off a local region of the universe and study it in isolation. The latter meant, very much against the spirit of Quantum theory, that the universe, including these building blocks, have properties and characteristics that are intrinsic, and independent of the observer. We saw in Chapter 3, how the EPR paradox was resolved in favour of Quantum theory by the experiment of Alain Aspect, and subsequent experiments. Interestingly, Aspect himself believes that his experiment debunks one of these two, locality and reality though he is not certain, which one!

These building blocks obey certain physical laws – the challenge of the physicists has been to discover and abstract these laws from the multitude of phenomena by observing the behavior of the building blocks. In a sense the laws so found are an a priori or handed down in advance feature of nature. For example Newton's laws or the laws of conservation of momentum or energy fall in this category. Newton did not invent them – they have been there from the beginning of time. He merely discovered them.

In contrast as noted above, there is the thermodynamic description in which there are a number of particles – these are typically molecules in different types of motion. We study in this system quantities like temperature, pressure and so on. These however are not fundamental in the sense of the velocity of a given particle. The temperature is proportional to the square of a velocity, but this velocity is not that of any particular particle. It is an average velocity. Indeed no real particle may have this actual velocity. In these considerations, we are not isolating the particles, but are studying their motion as a whole. The equations of thermodynamics (though based on the equations of mechanics) are therefore not fundamental in the sense of usual physics. Perhaps introducing the thermodynamic approach to fundamental physics could yield interesting results or even show a way out of the intractable problems confronting physics.

As Wheeler put it, [68], "All of physics in my view, will be seen someday to follow the pattern of thermodynamics and statistical mechanics, of regularity based on chaos, of "law without law". Specifically, I believe that everything is built higgledy-piggledy on the unpredictable outcomes of billions upon billions of elementary quantum phenomena, and that the laws and initial conditions of physics arise out of this chaos by the action of a regulating principle, the discovery and proper formulation of which is the number one task...."

In other words, perhaps there are no given in advance laws. Rather these emerge out of a background of "lawless", that is not regulated by any guiding a priori principle, chaotic events [21].

The reason this approach is more natural is, that otherwise we would be lead to ask, "from where have these laws come?" unless we either postulate a priori laws or we take shelter behind an anthropic argument. I remember attending a talk by a fairly well known cosmologist at a prestigious symposium in France. The speaker declared that the universe began as a star shaped object or something like that! You could legitimately ask, "Why a star, why not rugby ball shaped?" This is an extreme example, but it brings out the spirit of the earlier philosophy. On the other hand as we saw, the anthropic argument can be tricky, even slippery. This would tantamount to our saying, "Don't ask why the laws are so, or from where they have come. The plain truth is that if these laws had not been operating, there would be no sun, earth, origin of life on the earth, nor evolution, so that we would not exist to ask such a question."

An interesting but neglected body of work in the past few decades is that of Random or Stochastic Mechanics and Electrodynamics. To go back to our example of molecules in a container, if we try to specify the path of a particular particle over a few seconds, as we do in usual physics, it would show up as shown in (Cf.Fig.6.3).

To keep track of this complicated zigzag is a hopeless task, you will agree! And remember, there are so many other particles to reckon with. Studying all of them in terms of some minimum number of characteristics like temperature, pressure and so on, is not just practical, it is the only option. There is a long list of scholars who have contributed to these ideas over the years. They include names like De Pena, Timothy Boyer, Marshall, Nelson, Nielsen and so on. Though their work is interesting, particularly their attempts to deduce Quantum theory from these ideas, they have not been taken as seriously as they should have. In any case, I believe that the seeds of a new world view, of a paradigm shift are to be found here in the spirit of these considerations.

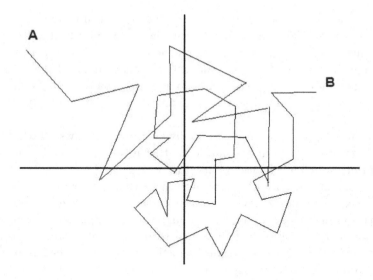

Fig. 6.3   Random motion of a molecule

In this context, I have proposed a variant model: that purely stochastic processes lead to minimum space-time intervals [69]. We can guess that such a minimum interval would be of the order of what is called the Compton wavelength and Compton time or, both put together the Compton scale. In one sense, this was noticed by Dirac himself following his deduction of the relativistic Quantum mechanical equation of the electron, which we encountered in the last Chapter. This equation also threw up several important predictions which were subsequently verified experimentally. You may remember these included the spin of the electron and also anti particles like positrons. But equally important Dirac noticed that his electron equation did not make any sense at all! It turned out that his electron had the velocity of light. Clearly that was not possible, because then it would have infinite mass and energy, apart from the fact that all this went against observation. We do see slow moving electrons everywhere. Dirac realized where the problem was. He argued that we are using point space and time intervals, that is whose length is zero. As he put it [70], "we can conclude that a measurement of a component of the velocity of a free electron is certain to lead to the result $\pm c$ (the speed of light). This conclusion is easily seen to hold also when there is a field present.

"Since electrons are observed in practice to have velocities considerably less than that of light, it would seem that we have here a contradiction with experiment. The contradiction is not real, though, since the theoretical velocity in the above conclusion is the velocity at one instant of time while observed velocities are always average velocities through appreciable time intervals."

To consider an example from our macro world, when we say that a car has been moving with a velocity of forty kilometers per hour, very strictly speaking, the velocity keeps fluctuating – it is only the average taken over several minutes which gives us the figure forty.

So firstly, there is this factor. Heisenberg's Uncertainty principle, though in the same spirit of knowledge and information through measurement, is slightly different. As Heisenberg emphasized, to measure we have to bombard the object, an electron let us say, with photons – light for example, which would disturb the electron's position in an unpredictable way.

In any case, Heisenberg's Uncertainty Principle forbids points in Quantum Mechanics – this would mean infinite momenta and energy which are physically meaningless, on the one hand. On the other hand to reemphasize, all this is physically meaningless because, in order to measure the momentum of a particle for example, we need to observe it for a length of time, to see that it has moved from one point to another. We can then determine its speed by dividing the distance covered by the time taken. The point is that some time and distance must be taken for all this. It cannot be done at one instant or one point. That would be meaningless.

So Dirac argued that our observation is over a small length of time, and then we are taking averages. Within this length of time we encounter what physicists call Zitterbewegung effects – that is a very rapid and unphysical vibration of the electron. Something resembling (Fig.6.4). The nett result is that we need to observe over a length of time and an interval in space. It is only the averages over these intervals that we use in physics as lengths or time intervals or speeds and so on. For example, the average vertical displacement over the horizontal segment OA in (Fig.6.4) is zero. On the whole the positives (above OA) cancel with the negatives (below OA). This scale is precisely the Compton scale. Why the Compton scale, you may ask. Because as we approach this scale any particle acquires a velocity which approaches c of light, which again is not meaningful.

The Compton scale modifies Einstein's special theory of relativity too. It can be argued in fact that within these intervals, special relativity does not

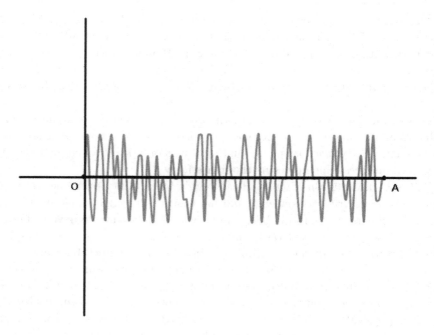

Fig. 6.4   Rapid unphysical fluctuations

hold good, including the neat cause and effect order! Weinberg notes [71], "Although the relativity of temporal order raises no problems for classical physics, it plays a profound role in quantum theories. The uncertainty principle tells us that when we specify that a particle is at position x_1 at time t_1, we cannot also define its velocity precisely. In consequence there is a certain chance of a particle getting from x_1 to x_2 even if (this has to be at a faster than light speed). To be more precise, the probability of a particle reaching x_2 if it starts at x_1 is nonnegligible (within the Compton scale) ... We are thus faced again with our paradox; if one observer sees a particle emitted at x_1, and absorbed at x_2... then a second observer may see the particle absorbed at x_2 at a time t_2 before the time t_1 it is emitted at x_1."

Most scientists, Weinberg included, try to rescue the situation. How can this be done? Remember the paradox is within the Compton scale. Imagine you were peering into this sample space through a magic telescope. Suppose a particle is created at A, then it darts across to B and gets destroyed there.

Now it is possible for another observer to see the particle being destroyed at B and then being born at A! Conventional wisdom dictates that the two observers see different particles. That way special relativity, so crucial for all of physics is saved. But then, as we have argued, how can we measure time itself, with our large devices, within these narrow confines, to make any statements at all?

To put it bluntly, spacetime points which are at the heart of physics, be it classical or Quantum, are physically meaningless. It is only in an idealized, what is called thermodynamic limit that we get spacetime points. This is the extreme case in which there are an infinite number of particles in the universe. Though the Heisenberg Principle in Quantum theory forbids arbitrarily small spacetime intervals, the above continuum character with space time points has been taken for granted even in Quantum Field Theory. In fact if we accept the proposition that what we know of the universe is a result of our measurement (which includes our perception), and that measurements are based on quantifiable units, then it becomes apparent that a continuum is at best an idealization.

We can understand this by considering a simple example. Suppose our units of time are minutes. Further, our digital clock displays the time directly. Then we can say that the clock is showing twelve. The next definable moment would be twelve-one. Now what does twelve mean? It means anything between twelve and twelve-one. Only when the digital dial shows twelve-one can we say that it is actually twelve-one. There is not much precision here. It doesn't matter if we have a smaller unit like a second or a nano second. Then twelve would mean an infinite number of instants between twelve and one second or one nano second past twelve. And so on. All this is behind the paradox of the point electron which was encountered in Dirac's Quantum mechanical treatment of the relativistic, spinning electron in which the electron showed up with the velocity of light, and infinite energies. This problem was encountered in classical theory as well.

Quantum mechanics has lived with a self contradiction: On the one hand we work with points of space and time, while on the other, the Heisenberg Uncertainty Principle prohibits meaningful physics for spacetime points. This as noted, is behind the troublesome infinite quantities we encounter. However, as we saw, we have bypassed these problems by resorting to mathematical devices.

Indeed it had been suggested by an American physicist, Hartland Snyder, Nobel Laureate T.D. Lee and others that the infinities which plague Quantum Field Theory are symptomatic of the fact that spacetime has a granular or discrete rather than continuous character. This has lead to a consideration of extended particles, as against point particles of conventional theory. Wheeler's spacetime foam and strings are in this class, with a minimum cut off at the Planck scale. As 't Hooft notes [72], "It is somewhat puzzling to the present author why the lattice (net like) structure of space and time had escaped attention from other investigators up till now...".

6.2 The Lawless Universe

All this has also lead to a review of the conventional concept of a rigid background spacetime. More recently, I have pointed out that it is possible to give a random and probabilistic underpinning to spacetime and physical laws. This is in the spirit of Wheeler's, "Law without Law" alluded to. In fact in a private communication to the author, Prof. Prigogine wrote, "...I agree with you that spacetime has a stochastic (probabilistic) underpinning".

Let us explore this line of thought a little more. Our starting point is the well known fact that in a random walk, as we saw, the average distance covered at a stretch is not the number of steps taken, N let us suppose, times the length of each step, one meter let us say, that is N meters. This is the case in our conventional space and time. Rather we must remember that sometimes as we saw, we go one step forwards and at times one step backwards. All this happens at random. Remember the experiment we performed? So sometimes, there could be two or three steps forwards before a few steps backwards are taken. The nett result is that the distance covered on the average is the square root of N. So, if we take a million steps this way, we would be a thousand meters or one kilometer from the starting point not a thousand kilometers. This distance could be backwards or forwards from our starting point. That is irrelevant. What matters is the distance from the starting point. This is very different compared to a "fixed" distance point of view.

The same game can be played with time as well. You might wonder how? We cannot go backwards in time. The whole point is that we allow time too to go backwards or forwards, though at the micro level – within the Compton scale, let us say. After all, even usual physics remains unchanged if time were reversed. What we encounter is really the final displacement in time – that is our unit of time. Such approaches were made in statistical mechanics and an American physicist, Edward Nelson mentioned as earlier, tried, though unsuccessfully to deduce Quantum theory on this basis, a few decades ago.

We get the same relation and situation in Wheeler's famous traveling problem and similar problems [73]. As the name suggests, in this scenario a salesman has to call on his various customers who are living in different cities. What he would like to do is expect to minimize the average distance he would have to travel per customer. This would save time, energy and cost. For example it would be very inefficient if he visited customer A and then customer D, bypassing customers B and C who are more or less on the route from A to D. He would then have to retrace much of his path to return to customers B or C (Cf.Fig.6.5).

Fig. 6.5   Travails of a traveling salesman

If the number of customers or cities to be visited is very large, in the hundreds, the number of ways he could visit all of them would be enormously large. Even a computer cannot compare all these possibilities and find the shortest distance for the salesman to cover all his customers. Nevertheless it is possible to conceive of what in statistical mechanics is called a mean free path, which is the average distance the salesman would have to cover to go from one customer to another. The Compton scale should be considered to be such a mean free path of statistical physics.

Let us consider the Compton wavelength l of a typical elementary particle and use a well known value N for the number of elementary particles. Suppose we are like the travelling salesman, going from one elementary particle to another using the distance l inbetween elementary particles. It turns out that if we travel a distance which is square root of N times the Compton length l, then we would have covered precisely the distance that we would in a Random Walk of N steps, as we just now saw. Miraculously this turns out to be the entire length of the universe itself!

From a different point of view, it is one of the cosmic "coincidences" or Large Number relations, pointed out by Hermann Weyl, the distinguished British physicist, Sir Arthur Eddington and others. In my work this relation which has been generally considered to be accidental (along with other such relations), is shown to arise quite naturally in a cosmological scheme based on fluctuations which are random events. They will be briefly discussed in the next Chapter. We would like to stress that we encounter the Compton wavelength as an important and fundamental minimum unit of length along with the even smaller Planck length. Such a minimum length makes the paths of particles irregular.

This irregular nature of the Quantum Mechanical path was noticed by Feynman [74] "...these irregularities are such that the 'average' square velocity does not exist, where we have used the classical analogue in referring to an 'average'.

"That is, the 'mean' square value of a velocity averaged over a short time interval is finite, but its value becomes larger as the interval becomes shorter. It appears that quantum-mechanical paths are very irregular. However, these irregularities average out over a reasonable length of time to produce a reasonable drift, or 'average' velocity, although for short intervals of time the 'average' value of the velocity is very high..."

It has to be pointed out that in the spirit of Wheeler's travelling salesman's "practical man's minimum" length, the Compton scale playing such a role, spacetime is like Richardson's delineation of a jagged coastline with a thick brush, the thickness of the brush strokes being comparable to the Compton scale. This thickness masks the irregularities, as we saw.

You may remember that Richardson found that the length of the common land boundaries claimed by Portugal and Spain as also Netherlands and Belgium, differed by as much as twenty percent! The answer to this non-existent border dispute lies in the fact that we are carrying over our concepts of smooth curves to the measurement of real life jagged boundaries or coastlines. As far as these latter are concerned, to re-emphasize, as Mandelbrot puts it "The result is most peculiar; coastline length turns out to be an elusive notion that slips between the fingers of one who wants to grasp it. All measurement methods ultimately lead to the conclusion that the typical coastline's length is very large and so ill determined that it is best considered infinite....." [75] The universe is like a Monet painting. I remember, in this connection, an amusing incident, some years ago. I was with a mathematician of international repute, having a cup of coffee in a restaurant in Bangalore. When I mentioned this to him, he was incredulous. He pulled out a paper napkin and started scribbling on it, muttering, this is impossible. After some fifteen minutes, he gave up, looking very perplexed indeed.

Spacetime, rather than being the smooth continuum of mathematicians, is more like a jagged coast line, a Brownian curve. In this case, a description in terms of points breaks down and the unit of measurement becomes crucial. All this has been recognized by some scholars, at least in spirit. As the Russian Nobel Prize winning physicist, V.L. Ginzburg puts it "The Special and General Relativity theory, non-relativistic Quantum Mechanics and present theory of Quantum Fields use the concept of continuous, essentially classical, space and time (a point of spacetime is described by four coordinates which may vary continuosly). But is this concept valid always? How can we be sure that on a "small scale" time and space do not become quite different, somehow fragmentized, discrete, quantized? This is by no means a novel question, the first to ask it was, apparently Riemann back in 1854 and it has repeatedly been discussed since that time. For instance, Einstein said in his well known lecture 'Geometry and Experience' in 1921: 'It is true that this proposed physical interpretation of geometry breaks down when applied immediately to spaces of submolecular order of magnitude. But nevertheless, even in questions as to the constitution of elementary particles, it retains part of its significance. For even when it is a question of describing the electrical elementary particles constituting matter, the attempt may still be made to ascribe physical meaning to those field concepts which have been physically defined for the purpose of describing the geometrical behavior of bodies which are large as compared with the molecule. Success alone can decide as to the justification of such an attempt, which postulates physical reality for the fundamental principles of Riemann's geometry outside of the domain of their physical definitions. It might possibly turn out that this extrapolation has no better warrant than the extrapolation of the concept of temperature to parts of a body of molecular order of magnitude'.

"This lucidly formulated question about the limits of applicability of the Riemannian geometry (that is, in fact macroscopic, or classical, geometric concepts) has not yet been answered. As we move to the field of increasingly high energies and, hence to "closer" collisions between various particles the scale of unexplored space regions becomes smaller. Now we may possibly state that the usual space relationships down to the distance of the order of $10^{-15} cm$ (Compton wavelength of proton like particles, a thousand trillionth of a centimeter) are valid, or more exactly, that their application does not lead to inconsistencies. It cannot be ruled out that, the limit is nonexistent but it is much more likely that there exists a fundamental (elementary) length $l_0 \leq$ (less than or equalling) $10^{-16} - 10^{-17} cm$ (a tenth of the above) which restricts the possibilities of classical, spatial description. Moreover, it seems reasonable to assume that the fundamental length l_0 is, at least, not less than the gravitational length (Planck length).

"... It is probable that the fundamental length would be a "cut-off" factor which is essential to the current quantum theory: a theory using a fundamental length automatically excludes divergent (infinite) results".

Einstein himself was aware of this possibility. As he observed, "... It has been pointed out that the introduction of a spacetime continuum may be considered as contrary to nature in view of the molecular structure of everything which happens on a small scale.... Then however, we must also give up, by principle the spacetime continuum. It is not unimaginable that human ingenuity will some day find methods which will make it possible to proceed along such a path. At present however, such a program looks like an attempt to breathe in empty space".

Nevertheless, Einstein himself was, in a sense, an extension of Plato and the Greek school of thought, for whom geometry was ultimate. He would say that a person not captivated by geometry was not born to be an inquirer into the secrets of nature. For this reason, he was not satisfied with his own equation of General Relativity. One side of the equation, expressing the geometry of the universe was pleasing. But the other side contained mechanical, not geometric quantities, which he did not like.

In contrast our space time given by distance R which is the diameter of the universe and its lifetime T represents a measure of spread – you can think of it as the maximum possible distance between two points but now in a slightly different sense. Remember the universe is a collection of N elementary particles, as seen earlier. We consider spacetime not as an apriori container of these particles but rather as a probabilistic collection of these particles, a random heap of particles. At this stage, we discard the concept of a continuum. We will return to this in the next Chapter but would like to point out that all this brings out apart from the random feature, a holistic feature in which the large scale universe and the micro world are inextricably tied up, as against the usual reductionist view. To better appreciate this, let us consider a random collection of N people and plot their IQ's. It is always found, provided the collection of individuals is random, that the number of people with extraordinarily high IQ (extreme right) is small as also those with high stupidity (extreme left) (Cf.Fig.6.6). Most of the people however have average intelligence plus, minus. R is the dispersion or maximum difference amongst people who are neither too stupid nor too intelligent whereas in the usual theory R would span the whole of the IQ axis. In our case, R stands for space or time extents.

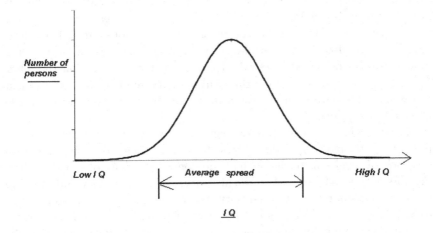

Fig. 6.6 IQ spread of people

There is another nuance. Newtonian space if you remember was a passive container which "contained" matter and interactions - these latter were actors performing on the fixed platform of space. But our view is in the spirit of Liebniz for whom the container of space was made up of the contents - the actors, as it were, made up the stage or platform. This also implies what is called background independence a feature shared by general relativity (but not string theory).

The view of spacetime in much of twentieth century physics can be compared to a smooth cloth, without any breaks or folds. This means that we could go down right up to a point of space or time. This is not a satisfactory state of affairs because Quantum theory itself as we saw declares that this is not very meaningful. Indeed all this is brought out by Quantum mechanical effect which we called Zitterbewegung, a German word literally meaning rapid vibration. If we go down to extremely small intervals of space or time we encounter exactly this feature where for example, the electron rapidly oscillates with the speed of light, which is meaningless. Moreover within these ultra small intervals, a sacred principle of science, that of causality, that is cause preceding effect also becomes dubious.

The discipline of Quantum Field Theory, that has been highly successful has managed to by pass these difficulties to a considerable extent as we saw. To put it simply the difficulties deal with infinite quantities which we encounter, for example for the mass or charge of a particle, when we go right down to a point of space or time. Such infinities are called divergences. Quantum Field Theory has devised the ingenious method or artifice called Renormalization that replaces the averaging that we just saw. Dirac himself was unconvinced and said, "I am inclined to suspect that the renormalization theory is something that will not survive in the future, and that the remarkable agreement between its results and experiments should be looked on as a fluke..."

Is nature crying out that our description is in some sense flawed? It is possible to by pass these difficulties, by taking them to mean that a description of nature in terms of the time honoured points of space or sharp instants of time is flawed. We saw this feature in string theory and other approaches like Loop Quantum Gravity. Such an approach can perhaps point to a solution of the decades old problem of providing a unified approach to gravitation and electromagnetism or to general relativity and Quantum theory. In fact string theory claims such a unification.

Einstein himself had toiled for fruitless decades in this direction and towards his last days he even lamented that he was nothing more than an old fool displayed on occasion and otherwise known for wearing shoes without socks. Richard Feynman too took a shot at this problem, but soon gave up, because, as he explained, there was no contact with experiment. That has been a persistent problem. As Wheeler notes, "no prediction of spacetime, therefore no meaning for spacetime is the verdict of the Quantum Principle. That object which is central to all of Classical General Relativity, the four dimensional spacetime geometry, simply does not exist, except in a classical approximation."

Let us recapitulate. The point is that apart from the Newtonian model of space which was the ancestor of later ideas, there was as we saw, an alternative idea of space put forward by Leibnitz, the German contemporary of Newton [52]. Not that space was an a priori container, but rather that the contents made up the container, that is space. This was a radically different approach and based on that today we can think of the scenario in the following manner. Space is not a platform on which the drama of nature is being enacted, the actors being matter, energy and forces. Rather as noted, it is these actors who constitute the platform. Clearly such a platform would be very different. For one thing it would not be a smooth fabric. It would be full of discontinuities or holes and kinks.

Next, in my approach nature, rather than being built on exact pre-given footrules and clocks and laws would be thermodynamic in the sense that these exact laws and footrules and clocks would be reduced to the status of temperature, which is related to an average velocity of the molecules of gas in a container. No single molecule might actually have this average velocity. Nevertheless the temperature represents some sort of a collective result of the movement of typically trillions of trillions of molecules.

So also our clocks, footrules and laws would become statistical rather than rigid, given in advance entities.

This leads to a picture where intervals smaller than a minimum can yield no meaningful physics. This would eliminate the infinities and would also remove what Wheeler described as the greatest crisis facing physics. He was referring to what happens at the very point and instant of the big bang. We encounter such, what are called singularities, inside black holes.

Now let us change our perspective. There is a special type of a black hole, one that has a charge and also keeps spinning like a top. This is called a Kerr-Newman black hole. Though this was supposed to be a huge object, let us see what happens if such a black hole were miniscule. Miraculously the Kerr-Newman black hole, provides a remarkably accurate description of the electron. Now the miracle here is to do with the fact that the spin of the electron is purely Quantum mechanical – it defies classical theory. On the other hand, the Kerr-Newman black hole is purely classical. Except of course for the spoilsport singularity where physics breaks down. However with the new thinking that points do not exist, the Kerr-Newman black hole can legitimately describe the electron and this would be the starting point of a meaningful description of the universe itself. Einstein would say, "It is enough to understand the electron..."

Intimately connected with the above thermodynamic approach is the ubiquitous Dark Energy, that has been observationally confirmed and reconfirmed since 1998. Think of it as a background disturbance, incessant and ubiquitous. We will come back to this new mystery again, but it now appears that the universe is immersed, as it were in a huge bathtub of Dark Energy.

My work over the past several years has been very much along these lines. I was lead to this thinking, because of the many loose ends in other approaches, ends which appeared un-tiable. We have encountered all these earlier. Particularly annoying are the infinities that crop up both in general relativity and Quantum theory. They could well point to the failure of theories based on spacetime points. So these points are fudged – spacetime is fuzzy. There is a minimum interval within which we have limited knowledge of positions, velocities and so on.

There is a nuance here. In some of the newer approaches, spacetime is like a net, a lattice. In this case however, if spacetime is fuzzy, then it cannot be like a sharp net either – such a net would at best be an approximate model. This minimum interval is the Compton scale alluded to above. Indeed within the Compton scale apart from the difficulties encountered by phenomena like Zitterbewegung or the rapid vibration, there is also the rather strange situation, as noted earlier that causality – the effect following the cause – becomes a casuality. Moreover as noted, how do we measure time inside the Compton wavelength to start with? We need a clock which consists of atoms and molecules all of which are much much larger than the Compton scale. So time within the Compton scale has no meaning.

A special case of the Compton scale is what we call the Planck scale – it is less than a billion, trillionth of the Compton scale and constitutes the smallest scale in the universe though this scale vanishes practically instantaneously. These constitute the limits of our knowledge, the limits to which we can meaningfully penetrate in order to retrieve useful information. The universe is immersed in a seething ocean of dark energy, continuously wriggling about in a chaotic manner. We will come back to this in the next Chapter but the point is that this background dark energy throws up the Compton scale including the Planck scale.

You can think of the dark energy in a simple model as vibrating springs. The size of the springs is at the Planck scale. Below the minimum scale the universe becomes totally chaotic – there is no meaningful universe to observe or speak about. But at this scale the springs can latch on to each other like hand holding school children and dance in unison. From such an array the particles begin to emerge at the Compton scale of the particles themselves. It must be noted that the tiny springs, if left to themselves, would evaporate off within the Planck time, in less than a trillion, trillion, trillionth of a second. But once they latch on to form particles – well these particles or these arrays of dark energy springs are at the lowest energy level and so are stable. It is rather like a person on the ground being stabler and safer than a person perched on a ladder. That is, the particles of the universe are long lived unlike their constituents, the Planck springs.

You could also think of particles as "droplets" that condense out of vapour. Next these particles interact and can constitute the entire universe of stars and galaxies. We have a similar situation in crystalline solids. The crystal mimics a series of atoms which are vibrating. The vibrations can get into unison and result in sound – this sound like electromagnetic radiation manifests itself as packets of sound or phonons. Particles are the counterpart of these phonons. The important point is that from a chaotic meaningless background dark energy, particles and the large scale universe itself emerge. We will return to these ideas in greater detail in the next Chapter.

There are several consequences of this model. The fact that the photon is to travel through the sea of dark energy means that there is a type of a viscous friction it encounters, like a ballbearing moving through a tube of oil. This friction results in a mass. We can also get this mass by considering the photon to be an array of Planck springs. This means that the photon which has been considered to be a massless object traveling with the speed of light has to be viewed like any other particle, but its mass is miniscule. Several experiments have been carried out in the past, using for example the magnetic field of the earth, to indicate that if the photon indeed has a mass, then this mass would have to be less than a certain limit – it would be less than a million, trillion, trillionth that of an electron. The mass of the photon that tumbles out of the above model is below this experimental limit. This means that it is meaningful. Moreover this tiniest of masses turns out to be same as the minimum mass in the universe that is allowed by Heisenberg's uncertainty principle itself. What an apparently miraculous coincidence! Looking at it another way, this may not be a coincidence at all.

Three American physicists, Alfonso Rueda, Bernhard Hirsch along with a third, H.E. Puthoff have also tried to explain mass in terms of a background viscosity. Their work was done in the pre dark energy days, but many have argued that this work reproduces renormalization.

Further, all this means that an elementary particle like an electron is not a point, because points have no meaning. Moreover, it is difficult to believe that different points, representing different types of particles, have complicated and differing properties. Now the electron can be modelled to be a small shell [76]. Such models for the electron have been considered for nearly a hundred years, and all turned out to contradict special relativity, apart from posing other problems. For example such an electron would not be stable at all. However all this was in an era in which there was smooth spacetime and no dark energy. Once we consider these new inputs, it can be argued that such a shell like electron is quite meaningful and stable.

We are considering here epoch turning concepts like spacetime resembling more a pock marked surface like that of the Moon or even a sponge, the photon having mass and so forth. The question is, though conventional theory has many shortcomings, does it mean that we should jettison these time honoured concepts? Again observation and experiment should play the role of the jury. Remember that Einstein's theory of relativity was formulated using the old concepts. So these new concepts would modify his theory of relativity. If the new ideas are meaningful then these modifications should be observed in real life. The catch has been that we need very high speeds, approaching that of light, and therefore very high energies. Fortunately there is an ingenious mechanism to test this. It is well known that the earth is bombarded with charged particles like protons as also radiation including very short wavelength potent radiation like gamma rays. Many of these are today known to originate in deep outer space. Collectively they are called Ultra High Energy Cosmic Rays. As they near the earth they collide with other particles resulting in the production of interesting byproducts. One mechanism by which this happens is called photo pion production, where pions are produced. If Einstein's theory is correct, then it was shown by three physicists, Greisen, Zatsepen and Kuzmin in the 1960s that protons of cosmic rays with energy greater than a certain limit cannot reach the earth. This energy is actually enormous. It is some trillion times that of the proton. So, if Einstein's special relativity is unmodified, then we should not be seeing cosmic rays with a greater energy than this limit. However there is preliminary evidence that this GZK limit has indeed been breached. For instance from the Akeno Giant Air Shower Array or AGASA in Japan at least twelve such events have been observed, which indicate that the GZK cut off has been breached.

Moreover if the photon has this miniscule mass, then gamma rays of different wavelengths or frequencies should move with different speeds. In the usual theory, all gamma rays move with the same speed, the speed of light. This difference in the speeds depends on the frequency or wavelength. This can be detected by gamma ray detectors. Already it is suggested that such differences or lags have indeed been observed. In the Summer of 2008 NASA shot into orbit the Gamma Ray Large Array Space Telescope or GLAST, which could also provide interesting clues of possible violations of Einstein's relativity theory. So there are indications, but as always in science, this preliminary evidence needs to be checked and crosschecked.

It may be mentioned that there are a few other independent approaches which also suggest similar effects, just as there have been other scholars who have also been working along the lines of a totally randomized substratum for the universe. Scholars like Amelino-Camelia of Rome have been working on what is sometimes called Double or Deformed Special Relativity in which the minimum Planck scale is introduced into Einstein's smooth spacetime.

However contra views are always viewed by the establishment with suspicion, which is good – and even derision, which is bad. This has been an age old story. A pupil of Pythagoras in ancient Greece had to commit suicide by drowning himself because he dared to suggest that there were numbers which could not be expressed as the ratio of two natural numbers or integers. Today such irrational numbers as mathematicians call them, once a heresy, are indispensable.

Modifications of special relativity is one aspect, but this could also leave its footprint on particle physics. The new ideas lead to a slight modification of the original Dirac equation for the electron via what has been called the Snyder-Sidharth Hamiltonian or energy [77]. The interesting thing about this new relation is that it could solve the puzzle of how the elementary particle masses are generated. You may remember that in the standard model, the Higgs mechanism had to be introduced to give the masses to the particles. But this mechanism has not been found in nature to-date. So perhaps we could bypass this difficulty. Equally interesting, this modified Dirac equation already has experimental backing. It tells us that the supposedly massless neutrino is actually massive [78]. As already noted, this indeed is the case, though it is sought to be explained in terms of a complicated mechanism.

Another feature of these ideas is that a unified description of gravitation and electromagnetism is possible, though this gravitation is a toned down version of general relativity. But then who knows, general relativity itself may not be the last word. In a paper written a few years ago, Nobel Laureate Martinus Veltman who was the supervisor of 't Hooft and shared the Nobel Prize with him, dared to express this opinion.

Following this train of thought too it is possible to obtain a simple formula that gives the masses of all known elementary particles, some two hundred and fifty of them to-date [79]. The formula is utterly simple. Take an integer, multiply it into a half integer, that is an odd number divided by two and multiply that into the mass of a pion. For different integers that we choose, we can get the masses of different elementary particles to a good degree of precision.

So diverse and disparate phenomena have their explanation in my approach and similar such approaches. Perhaps, an interesting and even satisfying aspect of all this is that there is an immediate route to conventional theory. The old theory turns out to be a special case of the new approach. In any case, to paraphrase what the famous Eddington once said, "If it looks like an apple, smells like an apple and tastes like an apple, then chances are that it may be an apple." Perhaps, the answers lie here, in these considerations.

Chapter 7

When the universe took a U turn

Out of the Unmanifest springs forth the Manifest

– Bhagawad Gita

7.1 The Exploding Universe

Albert Einstein stared at the photo plates incredulously, for a long time. And then, with a deep breath and a shake of his head he accepted the verdict. The universe was indeed flying apart.

Einstein's universe had resembled Newton's in some respects. The basic building blocks were stars, and then these stars were all scattered about and were stationary. This troubled Einstein – yes, Newton could explain why the moon doesn't crash down. But the stationary stars were another matter! They could be in equilibrium – but this would be like a pencil standing on its tip! The stars – the universe, would soon collapse. So Einstein was facing the problem that the Greeks did. Why do not all the stars collapse? He was forced to invent a repulsive force, cosmic repulsion, which counterbalanced gravitation and held the stars aloft.

Within a few years, in the 1920's however came two bolts from the blue. The first was a major discovery by the American astronomer Edwin P. Hubble. Using a recently installed giant telescope, largest at that time, he noticed that objects that had been dismissed as nebulae or small clouds, were anything but. They were teeming with stars, much like bacteria in murky water. Further, all the stars in a given nebula were almost at the same – and enormous – distance!

Suppose you looked around. You would see people around up to some meters away. Now you look through a telescope and see a huge group of people, all in the same direction and at the same distance. Only they are hundreds of kilometers away. You would immediately conclude that you were observing a distant town, and not just people in the vicinity. That is what Hubble had stumbled upon.

Not stars, but rather these huge conglomerations of stars or galaxies – star islands he called them – were the basic building blocks of the universe. He had transformed our vision of the universe. This momentous leap in vision did not in itself mitigate Einstein's problem of the collapse of the universe. It merely meant that the building blocks were galaxies. The next discovery did.

This was a few years after he had published his General Theory of Relativity in 1917. The photo plates he was now scrutinizing told a different story. These were the observations of V.M. Slipher. It turned out that these galaxies were all rushing away from each other. The universe was anything but still! It was exploding. There was no need for his invention – the counter balancing cosmic repulsion. Einstein would later call this his greatest blunder.

7.2   Birth Pangs of the Universe

Now what did this mean? Today the galaxies are farther away from each other than they were yesterday. Day before yesterday they would have been even closer. By simply back working, we could easily calculate that at a certain instant in the past, all of the matter in the universe was at a single point-almost [80]. The question is, how far back in the past? That depends on certain observational values, but today it appears that it could have been something like fifteen billion years ago.

This also means that at that instant, there was a titanic explosion, unmatched by even zillions of zillions of zillions of zillions of hydrogen bombs! This spewed out the matter which today surrounds us, and anything else besides. Astronomers have since called that event rather modestly, the Big Bang.

But have we been too simplistic in extrapolating what we see today into the remote past? What if the universe yesterday and the day before and so on was not much different from the universe today? After all scientists believe that the laws of the universe have remained pretty much the same. Then why not the universe itself? Such ideas were discussed by Thomas Gold, Hermann Bondi and Fred Hoyle many decades ago. As Sir Fred reminisced [81], "Meanwhile as early as 1947-48 a few of us in Cambridge were investigating a new physical idea in cosmology, namely that matter might be subject to a continuous form of creation. At first, Hermann Bondi would have none of it, although his close friend Tommy Gold was rather in favour of it. I was myself neutral to the idea. I realized in 1947, when Bondi and Gold turned to other ideas, that if continuous creation were to have any hope of acceptance it would have to be given a mathematical expression. In the latter part of 1947, I came to the conclusion that a new form of field would be needed, and that a scalar field was not only the simplest possibility but also the most promising. I wrote the field on paper as a capital C, and from then on it became known as the C-field. In January 1948 I found how to use the C-field in a modification of Einstein's equations with the result that the equations had as a particular solution what became later known as the Steady-State model. This, let me emphasize, was not a static model but one in which the main features of the universe are steady like a steadily flowing river. The universe expands but it does not become increasingly empty because new matter is constantly being created to make up the deficit produced by the expansion."

This was the explanation for the fact that the universe, at least as we would see it, didn't change with time, and yet at the same time the galaxies would be rushing away and disappearing into the remote recesses of space. The explanation was that as some galaxies hurtled away, they were replaced by brand new galaxies which had been created in the meantime. Created from nothing, or rather, something like nothing. You may object, "How can matter be created out of nothing?". So to put it another way, we could say that the matter for the new galaxies came from the same place or in the same way as the matter for the Big Bang! This theory, the steady state theory, was a serious competitor to the Big Bang Theory for sometime. Till the mid sixties to be precise.

At that time, a faint, barely discernible footprint of the Big Bang was accidentally discovered. For, at the time of the Big Bang, there would have been huge amounts of energy spewed around. George Gamow the celebrated American physicist discovered all this in a legendary paper of 1948. He had written the paper with a student, Ralph Alpher. The paper carries a third name though, that of the famous Nuclear physicist Hans Bethe. Now why did Gamow do that? He had a great sense of fun and could see that the names of the authors would now read, Alpher, Bethe, Gamow resembling "Alpha, Beta, Gamma", the first three letters of the Greek alphabet. Gamow felt that this was very appropriate for a paper about the beginning of the universe. In this paper, they observed that this initial energy would not simply vanish into nothing, but rather would get weaker and weaker with time and persist as some sort of a background radiation. Astronomers could even calculate the wavelength of this radiation today – it would be in the form of micro waves. They have a few millimeters as wavelength and we all use them in microwave ovens, in our kitchens.

Exactly such a cosmic microwave background radiation was discovered by Arno Penzias and Robert Wilson. Though they got the subsequent Nobel Prize, they had nothing to do with astronomy. In fact they were engineers working with the Bell Telephone Laboratories. The specific problem they were trying to overcome was that of eliminating disturbances from the ambient in a sensitive microwave detector. They were investigating if these disturbances in antennae could be caused by, for example, pigeon droppings! It is not everyday that one looks for bird droppings and finds instead the debris of the origin of the universe! The steady state theory was given a decent burial.

A recent variant of the steady state theory is the quasi Steady State model. This model originates in the ideas of Arp a defiant astrophysicist from a Max Planck Institute in Germany, Narlikar the former Indian student of Fred Hoyle, Hoyle himself, and others. In this theory, there is not one Big Bang, but rather, a series of mini Big Bangs taking place all the time and spread across the universe. In each of these mini Big Bangs matter equivalent to hundreds of thousands of trillions of solar masses is created. The model explains a number of observational features, but remains a minority view.

7.3   The Infant Universe

By now physicists had realized that the very early universe would be quite unlike anything astronomers could have envisaged in the past. This is because at the very beginning of the universe, almost at the instant of the Big Bang, the universe would be smaller than the Planck length, that is one billion, trillion, trillionth of a centimeter. This would be the case, approximately one million, trillion, trillion, trillionth of a second after the Big Bang. You may wonder why we have excluded the instant of the Big Bang. This is because as you may remember, that very instant represents a singularity, an instant at which there is no meaningful physics to speak of. Or we could say that there is nothing of any meaning at this time. So coming back and picking up threads, we are talking about a time when the temperature of the universe would be some trillion, trillion, trillion degrees centigrade. But notice that at this micro miniscule size we are in the domain of Quantum Gravity – where we believe that gravitation joins the other three forces and is indistinguishable from them, the point of unification. This means that we have to jettison our large scale cosmology of the universe of galaxies and return to Quantum considerations.

So we are at an instant where gravity separates out from the other three forces, as the universe expands and cools. One way of understanding this is by performing an experiment with water. If water is at room temperature, in a sense there is total randomness of the jiggling molecules, as we saw earlier. Now let us cool water. As we reach the freezing point, around zero degrees, suddenly the random collection of molecules constituting liquid water freeze into solid ice, which is very different in character. Ice is a series of ordered crystals, each crystal having a characteristic hexagonal shape. We could say that from a state of (liquid) total randomness we have, in a very short time, around the freezing temperature transition into a state where the molecules are highly ordered. Physicists call such a rapid transition, a phase transition. A similar thing happened in the very early universe – gravitation separated out from the other three forces, in a phase transition.

Meanwhile the expansion continued to cool the universe and the strong force separated from the electro weak force. When the expansion and cooling brought the universe to a size of something less than a football, this being a billionth of a second after creation, the electro weak force separated into our familiar electromagnetic and weak forces. At that stage the temperature of the universe was some ten thousand trillion degrees. So finally at this football sized stage of the universe, a billionth of a second after creation, all the four forces familiar to us today came into play. Nevertheless this was still a time when in a sense the universe was a hot soup consisting of quarks, electrons and photons, the basic constituents of just about everything.

The explosion kept blowing the universe apart meanwhile, and the quarks combined to form protons, neutrons, pi-mesons and so on. It was this type of a messy soup which George Gamow originally had in mind and christened as 'Ylem', the original Greek conception of primordial matter. Remember the universe was still just three minutes old [82] This was the stage when stable nuclei, needed for atoms began to form. However it still took some three hundred thousand years for the first atoms to emerge –

these atoms needed temperatures of a few thousand degrees at most to exist in a stable form. It was at this stage that radiation including light could traverse large distances. The stable atoms were still hurled about by the continuing expansion, but managed to accumulate or aggregate into clumps, later galaxies, clusters and so on.

7.4   A Dark Mystery

Remember that we have completed only half the story-from the beginning to the present day. What happens next? Would the universe continue to expand and expand for ever and ever and end up, as practically nothing? Or would the expansion halt and revert into a collapse? After all, astronomers argued, there is no force pushing the universe outwards. It is a bit like a stone that is thrown up. It rushes higher and higher, slowing down all the while, till it stops before hurtling back. This would also mean, a collapse to a single point or something thereabouts, triggering off the next Big Bang. And so on and so on.

Astronomers need to know the density of the universe to answer this question. To put it simply, if there is enough material content in the universe whose gravity would drag the expansion due to the initial explosion, to a halt, then we would have an oscillating type of a universe. If not, the universe would blow out into nothingness. Remember the escape velocity, the velocity needed to escape from a planet's gravity, for example. The more massive the planet, the greater is the escape velocity, the more difficult it is to escape.

Prima facie, it appeared that the material content of the universe was much less than required to halt and reverse the expansion. But there have been contra indicators. We would expect that the velocities of the stars at the edge of the galaxies would drop off. This does not appear to be the case. The speeds at which galaxies rotate do not match their material content. Astronomers have suspected that there is more than meets the eye – some hidden matter, or dark matter which we are not able to account for. This was postulated as long back as the 1930s to explain the fact that the velocity curves of the stars in the galaxies did not fall off, as they should. Instead they flattened out, suggesting that the galaxies contained some undetected and therefore non-luminous or dark matter.

The identity of this dark matter is anybody's guess though. It could consist of what particle physicists call Weakly Interacting Massive Particles

(WIMPS) or the super symmetric partners of existing particles. Or heavy neutrinos or monopoles or unobserved brown dwarf stars and so on. In fact Prof. Abdus Salam speculated some two decades ago [42]. "And now we come upon the question of dark matter which is one of the open problems of cosmology. This is a problem which was speculated upon by Zwicky fifty years ago. He showed that visible matter of the mass of the galaxies in the Coma cluster was inadequate to keep the galactic cluster bound. Oort claimed that the mass necessary to keep our own galaxy together was at least three times that concentrated into observable stars. And this in turn has emerged as a central problem of cosmology.

"You see there is the matter which we see in our galaxy. This is what we suspect from the spiral character of the galaxy keeping it together. And there is dark matter which is not seen at all by any means whatsoever. Now the question is what does the dark matter consist of? This is what we suspect should be there to keep the galaxy bound. And so three times the mass of the matter here in our galaxy should be around in the form of the invisible matter. This is one of the speculations."

Add this dark matter, and who knows? We may just about have enough matter to halt the explosion. That has been the generally accepted view for many years. The exact nature of dark matter itself, has as noted eluded astronomers.

There still were several subtler problems to be addressed. One was the famous horizon problem. To put it simply, the Big Bang was an uncontrolled or random event and so, different parts of the universe in different directions were disconnected at the very earliest stage and even today, light would not have had enough time to connect them. So they need not be the same. Observation however shows that the universe is by and large uniform, rather like people in different countries exhibiting the same habits or dress. That would not be possible without some form of faster than light intercommunication which would violate Einstein's special theory of relativity.

The next problem was that according to Einstein, due to the material content in the universe, space should be curved whereas the universe appears to be flat. There were other problems as well. For example astronomers predicted that there should be monopoles that is, simply put, either only North magnetic poles or only South magnetic poles, unlike the North South combined magnetic poles we encounter. Such monopoles have failed to show up even after seventy five years as noted earlier.

Some of these problems, were sought to be explained by what has been called inflationary cosmology whereby, early on, just after the Big Bang the explosion was super fast [83]. To understand this let us go back to cooling water again. But this time we do it so carefully that even when we reach the freezing point, we do not allow the water to solidify. Rather we continue to cool the water below zero, ensuring that it is still a liquid. This, by the way, is possible. Now such water is called supercooled water, but it is very unstable as you can guess. The slightest disturbance and it solidifies double quick. Inflation is such a super quick process.

What would happen in this case is, that different parts of the universe, which could not be accessible by light, would now get connected because of this super fast expansion. At the same time, this super fast expansion in the initial stages would smoothen out any distortion or curvature effects in space, leading to a flat Universe and in the process also eliminate the monopoles.

Nevertheless, inflation theory has its problems. It does not seem to explain the cosmological constant or cosmic acceleration observed since. Further, this theory seems to imply that the fluctuations it produces should continue to indefinite distances. Observation seems to imply the contrary.

Several times in the past, as we saw, everything had fallen into place. Almost. The Greek model for instance. Or, for another example, at the turn of the last century, classical physics had all the answers, almost. In fact in 1928, Max Born had declared that physics would end after six months. His comment was inspired by Dirac's just discovered equation of the electron – the "last" missing piece in the puzzle.

7.5   The U turn

Around this time, in 1997 I had put forward a radically different model. In this, there wasn't any Big Bang, with matter and energy being created instantaneously. Rather the universe is permeated by an energy field of a kind familiar to modern physicists. The point is, that according to Quantum theory which is undoubtedly one of the great intellectual triumphs of the twentieth century, all our measurements, and that includes measurements of energy, are at best approximate. There is always a residual error. This leads to what physicists call a ubiquitous Zero Point Field or Quantum Vacuum. We will return to this "Dark Energy" soon. Out of such a ghost background or all pervading energy field, particles are created in a totally random manner, a process that keeps continuing. However, much of the matter was created in a fraction of a second. There is no "Big Bang" singularity, though, which had posed Wheeler's greatest problem of physics. The contents of this paper went diametrically opposite to accepted ideas, that the universe, dominated by dark matter was actually decelarating. Rather, driven by dark energy, the universe would be expanding and accelerating, though slowly. I was quite sure that this paper would be rejected outright by any reputable scientific journal. So I presented these ideas at the prestigious Marcell Grossmann meet in Jerusalem and another International Conference on Quantum Physics. But, not giving into pessimism, I shot off the paper to a standard International journal, anyway. To my great surprise, it was accepted immediately!

There is a further cosmic foot print of this model: a residual miniscule energy in the Cosmic Microwave Background, less than a billion billion billion billionth of the energy of an electron. Latest data has confirmed the presence of such an energy. All this is in the spirit of the manifest universe springing out of an unmanifest background, as described in the Bhagvad Gita. There are several interesting consequences.

Firstly it is possible to theoretically estimate the size and age of the universe and also deduce a number of very interesting interrelationships between several physical quantities like the charge of the electron, the mass of elementary particles, the gravitational constant, the number of particles in the universe and so on. One such, connecting the gravitational constant and the mass of an elementary particle with the expansion of the universe was dubbed as inexplicable by Nobel Laureate Steven Weinberg. But on the whole these intriguing interrelationships have been considered by most scientists to be miraculous coincidences.

With one exception. The well known Nobel Prize winning physicist Paul Dirac sought to find an underlying reason to explain what would otherwise pass off as a series of inexplicable accidents. In this model, there is a departure from previous theories including the fact that some supposedly constant quantities like the universal constant of gravitation are actually varying very slowly with time. Interestingly latest observations seem to point the finger in this direction.

However my model is somewhat different and deduces these mysterious relations. Further, it sticks its neck out in predicting that the universe is not only expanding, but also accelerating as it does so. This went against all known wisdom. Shortly thereafter from 1998 astronomers like Perlmutter and Kirshner began to publish observations which confirmed exactly such a behavior [84]. These shocking results have since been reconfirmed. The universe had taken a U Turn.

When questioned several astronomers in 1998 confided to me that the observations were wrong! After the expansion was reconfirmed, some became cautious. Let us wait and see. At the same time, some rushed back to their desks and tried to rework their calculations. The other matter was, what force could cause the accelerated expansion? The answer would be, some new and inexplicable form of energy, as suggested by me. Dark Energy. Later the presence of dark energy was confirmed by the Wilkinson Microwave Probe (WMAP) and the Sloane Digital Sky Survey. Both these findings were declared by the prestigious journal Science as breakthroughs of the respective years.

The accelerated expansion of the universe and the possibility that supposedly eternally constant quantities could vary, has been the new paradigm gifted to science, a parting gift by the departing millennium.

A 2000 article in the Scientific American observed, "In recent years the field of cosmology has gone through a radical upheaval. New discoveries have challenged long held theories about the evolution of the Universe... Now that observers have made a strong case for cosmic acceleration, theorists must explain it.... If the recent turmoil is anything to go by, we had better keep our options open."

On the other hand, an article in Physics World in the same year noted , "A revolution is taking place in cosmology. New ideas are usurping traditional notions about the composition of the Universe, the relationship between geometry and destiny, and Einstein's greatest blunder."

7.6   A Darker Mystery

In a sense all this resurrects the cosmic repulsion which Einstein mistakenly introduced around 1920, and later characterized as the greatest blunder of his life. But it also raises the question, what is the source of this repulsion? Scientists are again looking up to the mysterious force of the Quantum Vacuum for an answer. In one version it is called quintessence. Suddenly the scientific community has switched from swearing by dark matter to invoking dark energy.

In fact the concept of a Zero Point Field (ZPF) or Quantum vacuum (or Aether) is an idea whose origin can be traced back to Max Planck himself. Quantum theory attributes the ZPF to the virtual Quantum fluctuations of an already present electromagnetic field. What is the mysterious energy of supposedly empty vacuum?

It may sound contradictory to attribute energy or density to the vacuum. After all vacuum in the older concept is a total void. However, over the past four hundred years, it has been realized that it may be necessary to replace the vacuum by a convenient medium chosen to suit the specific requirements of the time. For instance Descartes the seventeenth century French philosopher mathematician proclaimed that the so called empty space above the mercury column in a Torricelli tube, that is, what is called the Torricelli vacuum, is not a vacuum at all. Rather, he said, it was something which was neither mercury nor air, something he called aether.

The seventeenth century Dutch Physicist, Christian Huygens required such a non intrusive medium like aether, so that light waves could propagate through it, rather like the ripples on the surface of a pond. This was the luminiferous aether. In the nineteenth century the aether was reinvoked. Firstly in a very intuitive way Faraday could conceive of magnetic effects in a convenient vacuum in connection with his experiments. Based on this, the aether was used for the propagation of electromagnetic waves in Maxwell's theory of electromagnetism, which in fact laid the stage for special relativity as we saw.

This aether was a homogenous, invariable, non-intrusive, material medium which could be used as an absolute frame of reference at least for certain chosen observers. The experiments of Michelson and Morley towards the end of the nineteenth century which we encountered earlier were sought to be explained in terms of aether that was dragged by bodies moving in it. Such explanations were untenable and eventually lead to its downfall, and thus was born Einstein's special theory of relativity in which there is no such absolute frame of reference. The aether used like a joker in the pack lay discarded once again.

Very shortly thereafter the advent of Quantum mechanics lead to its rebirth in a new and unexpected avatar. Essentially there were two new ingredients in what is today called the Quantum vacuum. The first was a realization that classical physics had allowed an assumption to slip in unnoticed: In a source or charge free "vacuum", one solution of Maxwell's equations of electromagnetic radiation is no doubt the zero or nothing solution. But there is also a more realistic non zero solution. That is, the electromagnetic radiation does not necessarily vanish in empty space.

The second ingredient was the mysterious prescription of Quantum mechanics, the Heisenberg Uncertainty Principle, according to which as we saw, it would be impossible to precisely assign momentum and energy on the one hand and spacetime location on the other. Clearly the location of a vacuum with no energy or momentum cannot be specified. There would be areas where energy would emerge from nothing and elsewhere disappear into nothing. This need not go against the hoary Law of Conservation of Energy, for on the average, the appearing and disappearing energy could cancel out.

This leads to what is called a Zero Point Field. For instance, a swinging pendulum, according to classical ideas has zero energy and momentum in its lowest position or before it starts swinging. But according to the Heisenberg Uncertainty Principle a subatomic pendulum would be vibrating. You may remember the Planck springs we encountered in the last Chapter. Imagine beads attached at their ends. Classically you could keep them motionless in a vacuum. Now they are oscillating. You could say that they are being buffeted randomly by the Zero Point Field. At this stage there is no physics. But occasionally they could vibrate in unison. Then physics indeed the universe itself would be born, (Cf.Fig.7.1) out of this dance of the oscillators. This provides an understanding of the fluctuating electromagnetic field in vacuum.

From another point of view, according to classical ideas, at the absolute zero of temperature, there should not be any motion. After all the zero is when all motion freezes. But as Nernst, father of this third law of thermodynamics himself noted, experimentally this is not so. There is the well known super-fluidity due to Quantum mechanical – and not classical – effects. This is the situation where supercooled Helium moves in a spooky fashion rather than stay put as it is supposed to do.

This mysterious Zero Point Field or Quantum vacuum energy has since been experimentally confirmed in effects like the Casimir effect which demonstrates a force between uncharged parallel plates separated by a charge free medium, the Lamb shift which demonstrates a minute jiggling of an electron orbiting the nucleus in an atom – as if it was being buffeted by the Zero Point Field, and so on.

The Quantum vacuum is a far cry however, from the passive aether of olden days. It is a violent medium in which charged particles like electrons and positrons are constantly being created and destroyed, almost instantly, in fact within the limits permitted by the Heisenberg Uncertainty Principle for the violation of energy conservation. One might call the Quantum vacuum as a new state of matter, a compromise between something and nothingness.

(a) Stationary spring in a classical vacuum

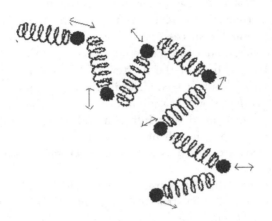

(b) The random spring in quantum vacuum

(c) Springs oscillate in unison

Fig. 7.1   Planck "Springs" in the quantum vacuum

Something which corresponds to what the Rig Veda described thousands of years ago: "Neither existence, nor non existence."

In any case a Quantum vacuum can be a source of cosmic repulsion. However a difficulty in this approach has been that the value of the repulsion or cosmological constant turns out to be huge, far beyond what is observed. In fact going by this, as noted earlier the universe should have blown out long long ago. This has been called the cosmological constant problem.

My approach has been that the energy of the fluctuations in the background vacuum electromagnetic field could lead to the formation of elementary particles. Indeed in another era, this was Einstein's belief. In the words of the American Nobel Laureate Wilzeck [85], "Einstein wanted to regard the fields, or ethers, as primary. In his later work, he tried to find a unified field theory, in which electrons (and of course protons, and all other particles) would emerge as solutions in which energy was especially concentrated, perhaps as singularities. But his efforts in this direction did not lead to any tangible success."

In some sense, this model bears a resemblance to both the Big Bang and Inflation Theories. Much of the matter of the universe would still have been created within the first second, while the repulsion and flatness of the universe are ensured. However the majority view is that the standard Big Bang Theory as it is called with Inflation and recent modifications seems to hold out some hope, though many question marks persist. Is this the Greek Epicycle syndrome? It is not easy, as we saw, to break away from the past.

Another iconoclastic dramatic observation which is gaining confirmation is that, what is called the fine structure constant, which scientists have considered to be a sacrosanct constant of the universe, has in fact been slowly decreasing over billions of years. Webb and his co-workers in Australia have confirmed this by observing the spectrum of light from the distant Quasars and comparing this with spectra in the vicinity. As the fine structure constant is made up of the electric charge of the electron, the speed of light, and what we called the Planck constant, this would mean that one or some or even all of them are not the sacred constants they have been taken for, but are slowly changing with time. The consequences of this could be dramatic. For instance this would mean that atoms and molecules in the past were not the same as their counterparts today-this will be true in future too. Some are questioning the constancy of the sacrosanct speed of light, and even Einstein's theories, themselves. Perhaps some of these older ideas are very good approximations – indeed Newton's laws still are – but not the exact or full story.

7.7   Other Footprints

This new dark energy driven fluctuational cosmology predicts that the sacrosanct constant of gravitation G, introduced first by Newton actually decreases slowly with time. Such ideas have been put forward before. There is a cosmology which goes after the names of its American inventors, Brans and Dick. So also there is the cosmology introduced by Dirac. *G* changes with time in these theories, though a little differently. These cosmologies have not gone very far, as they soon came up with contradictions. The crucial point is, exactly how does the constant of gravitation *G* change with time, and if so what are the observable effects. In our case, it can be argued that this variation gives an explanation of almost all the peculiar effects of general relativity. Let us quickly see what these are.

Firstly there is what is called the precesion of the perihelion of the planet Mercury. We all know that Mercury's orbit is quite elliptical in comparison with many other orbits, for example that of the Earth. The perihelion is the point where Mercury comes closest to the Sun (Cf.Fig.7.2). You might expect that this point is fixed in space. Actually it moves backwards, though by a very minute amount, less than one sixtieth of a degree per century. This could not be fully explained by Newton's gravitational theory, but it can be explained by general relativity.

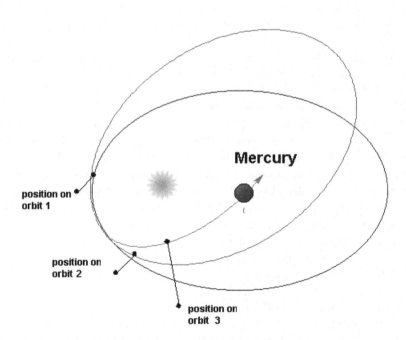

Fig. 7.2   Precession of perihelion of planet Mercury

We have already seen how light is bent by gravitation, for example the light from a star as it grazes past the Sun on its way towards us. You may remember that this was the first test of general relativity that was observationally confirmed by Sir Arthur Eddington, way back in 1919.

There were subtler effects which have been discovered more recently. For instance there are several examples of a pair of stars which are made up almost completely of neutrons. These are neutron stars. Incidentally pulsars which emit discrete radio pulses are believed to be neutron stars. So we could have a pair of neutron stars orbiting each other like partners on a ball room dance floor. Astronomers have found that the period with which this pair of stars circle each other keeps decreasing by a very minute amount. This can be ingeniously explained by general relativity. When such massive objects gyrate, they set off oscillations in spacetime itself. These oscillations are like ripples on the surface of a pond, except that the surface is spacetime itself. Such ripples are called gravitational waves, and they carry away energy. Though gravitational waves themselves have not been detected directly, in the case of the binary or pair of pulsars, the gravitational waves could explain the decrease in the time period. As the gravitational waves cart off energy, the pulsar pair loses this energy. This leads to the change in the period of orbit.

Interestingly all these effects can also be the result of a decrease with time in the gravitational constant G. Moreover such a decrease could also be invoked to explain the otherwise inexplicable acceleration of a couple of space crafts. Pioneer 10 and Pioneer 11 were launched by NASA a few decades ago. The space crafts are now leaving the solar system, but are also leaving behind a mystery as they depart. There is a very tiny acceleration they are exhibiting, for which there is no satisfactory explanation. The decrease with time of the gravitational constant could be the reason. Equivalently, this mimics the acceleration of the universe itself as it is egged in by dark energy. Finally such a decrease of G leaves a very tiny footprint in that the orbital distance of planets from the sun would change by a miniscule amount, as indeed the sizes of massive objects like galaxies. This is an effect which is yet to be confirmed conclusively, but in any case it is not predicted by general relativity.

All this apart, there are a number of relations which involve the number N of elementary particles in the universe and other measurable quantities like the radius of the universe, the velocity of light, the mass of the pion, the gravitational constant G and so on. One of the other, what may be called large scale parameters, is the Hubble constant H. You may remember the discovery of the expanding universe. At that time Edwin Hubble discovered how the universe explodes. The farther the galaxies, the faster they rush out. That is, the velocity of the galaxy equals the Hubble constant H times its distance. Indeed we have encountered one of these relations in our discussions about the Random Walk. This is the Weyl-Eddington relation referred to earlier. The size of the universe is square root of N times the pion Compton wavelength. This relation is undoubtedly a marvel, but it doesn't pose any philosophical problem in the sense that a large scale parameter equals another large scale parameter. However there is another relation between the mass of the pion m, and the Hubble constant which is very puzzling. This is because, in our usual reductionist approach, the mass of the pion is a very local property – it is like a building block which has nothing to do with a structure that emerges out of the building blocks. But then it is related to the Hubble constant which is a large scale parameter, that is it is directly related to the super structure. This makes the relation even more puzzling. As Weinberg notes [71], "...it should be noted that the particular combination of $\hbar$, (*the Planck constant*)H, G, and c appearing (in the formula) is very much closer to a typical elementary particle mass than other random combinations of these quantities; for instance, from $\hbar$, G, and c alone one can form a single quantity ... with the dimensions of a mass, but this has the value (one hundred thousandth of a gram) more than a typical particle mass by about 20 orders of magnitude (that is nearly a billion trillion!)

"In considering the possible interpretations (of the formula), one should be careful to distinguish it from other numerical "coincidences"... In contrast, (the formula) relates a single cosmological parameter, H, to the fundamental constants $\hbar$, G, c and m, and is so far unexplained."

It is extremely unlikely, though not impossible that such relationships should all be coincidences or accidents. These would be remarkable, marvelous accidents. In one sense accepting such coincidences is more like elevating them to the status of miracles. It goes against the spirit of scientific enquiry.

Dirac however took the minority view that these various coincidental relations are telling us something, that there is some underlying theory, which he tried to uncover. Though he tried for many years, he failed to come up with a cosmology that was consistent.

In contrast, in the above cosmology of fluctuations, all these "miraculous" relations are no longer ad hoc accidental marvels, but rather are a result of the underlying theory. In other words they are part of the physics of how the universe is born and evolves. There is here a parallel with the old Greek universe. Suppose you built a clock work with pulleys and all that to illustrate the Greek model. You could do the same to illustrate the Newtonian universe – in fact such a model of the solar system is an exhibit in many Science Museums. An important difference between the two would be [86] that in the former model you would have to have several cranks to set the different individual spheres into motion, whereas in the latter model you would just need one crank. So several ad hoc Greek rotations were replaced by one theory of universal gravitation.

Economy of hypothesis. That is a very important guideline for science which points towards truth. The unification that we examined so closely is very much in this spirit. The model of fluctuations and particle creation from the dark energy sea explains so many disparate relations, as a byproduct It also answers Weinberg's troublesome question – how can the mass of a pion, which is purely a local characteristic or building block depend on the large scale cosmos. In fact this model argues that the microscopic scale and the large scale, that is the building blocks and the structure are intertwined. This is very much in the spirit of the philosophy of the nineteenth century scholar, Ernst Mach. Mach argued that, for instance the mass of a particle has no meaning if the particle were the only object of the universe. The mass is rooted in the rest of the stars around us. We can understand this argument by considering a charged particle like an electron. If this were the only particle in the universe, its charge would be meaningless. We would need at least another charge to make the electron's charge to have any sense at all. Einstein admired Mach's ideas, but didn't actually use them anywhere. (I would go one step further. The charge is so fundamental a concept, that it is more a property of spacetime, rather than being anything separate.) This new cosmology explains another puzzle too. It now appears that there is a small acceleration present all across the universe. Such an acceleration is not confined to the Pioneer spacecrafts alone. This is a puzzle, but can be explained in terms of the ubiquitous dark energy.

7.8 A New Dawn

At the other end of physics, is the standard model of particle physics which we encountered earlier. This attempts to explain what are called strong nuclear forces (which hold particles like protons, neutrons and even the nucleus itself together) and weak forces (which cause radioactive decay) and finally the familiar electromagnetic forces, all in a single unified scheme. However latest developments have caused some further tinkering of this standard model. One such development was the discovery in the late nineties that the neutrino, an elusive and supposedly massless particle, has a mass, negligible though it is. The standard model you may remember requires the neutrinos to be massless. Then the fact remains that this model relies as we saw, on eighteen different quantities which have to be specified ab initio. This is almost like the model of a God one each for each and every phenomenon, as in pre historic times. Any number of them. Eighteen plus at the end of the second millennium. Some progress. And there are other loose ends. As we saw, the standard model requires particles called the Higgs bosons. They have eluded detection from the 1960's. The hope is that these truant particles will be detected by CERN's latest wonder gadget, the Large Hadron Collider. This hope has been dampened a bit, in recent times. This is because, the older American Bevatron accelerator now seems to be ruling out a Higgs mass in the higher range. The LHC it was hoped would be effective at the higher energies. But the last word hasn't been said.

So where are we at the dawn of a new millennium? To be precise we are in the midst of an irreconcilable marriage between the ultra small and the macro large. For, the twentieth century threw up the two great intellectual pillars: general relativity as applied to the universe at large, and Quantum theory as applied to the sub atomic world. Or, put it another way, the force of gravitation of the universe at large and the other forces of the atoms and nuclei. Both of these have been great triumphs, but in their own domains. The greatest failure has been that the two have not dovetailed. It is almost as if the universe is schizophrenic, as if there are two parallel universes, the large and the small. The problem is that general relativity has some unique characteristics, like the curvature of space and so does Quantum theory, for example the peculiar spin of an electron. But these characteristics do not blend.

If we look back, Newton's universe was one in which space was like an absolute background, a stage on which actors like matter and energy played their roles in time, which was a totally independent entity. After more than two hundred years, Albert Einstein brought about a great unification – he fused space and time in his special theory of relativity. Nevertheless, what was now spacetime was the stage on which the drama of the universe was being enacted.

Some ten years later, Einstein brought about another fusion, in his general theory of relativity: He combined the stage of space time with an actor, namely the gravitational force which controls the large scale cosmos. The stage had now become an actor. It became dynamic, describing gravitation. All this cannot however be combined with Quantum theory or sub atomic forces. Several fruitless decades have gone by in such an attempt. For, it is difficult to believe that the universe is schizophrenic. Einstein himself spent the last thirty years of his life in this futile endeavor. As noted, he lamented, in later years, "I have become a lonely old chap, who is mainly known because he doesn't wear socks and who is exhibited as a curiosity on special occasions".

Interestingly though, both Quantum theory and general relativity share one characteristic as we have seen. Their spacetime can go down to intervals as small as we please, right up to a point in fact.

Recent models you may remember have questioned this tacit assumption, which has slipped in, almost as if it has been undetected.

Suppose as we discussed, we cannot talk of spacetime points or instants— suppose we can only go up to a minimum interval and no further? Suppose what happens within these minimum interval is not that clear or specific? This leads to several interesting possibilities. Such possibilities are being investigated in several recent models such as Quantum super strings or Loop Quantum Gravity and my own model of fuzzy spacetime. It now becomes possible to reconcile general relativity and Quantum theory at least to a certain extent.

Though Quantum super string theory holds out promise, there is a price to pay. Several extra dimensions for the universe have to be invoked, though at the end of the day it is explained that the extra dimensions somehow curl up and vanish. Even in the old Einsteinian days, it was known that electromagnetism and gravitation could be combined by invoking extra dimensions but that didn't work. I remember sitting beside Eugene Wigner at a Symposium, in the early days of the second generation of extra dimensions in string theory. A distinguished physicist was explaining how extra dimension would solve certain problems. Prof. Wigner stood up, "Excuse me, how does this work?" The physicist explained with great caution and respect that extra dimensions are being used. "What do these dimensions mean?" Professor Wigner asked, "I do not understand all this". And he sat down, thoroughly dissatisfied.

Another difficulty is that a verification of many of these ideas requires huge particle accelerators which are beyond the capability of today's technology and budgets. Who knows how long we have to wait? From this point of view, namely that there is a lack of verifiability, some even question whether Quantum super string theory is a theory at all. We saw all this earlier on. My model on the other hand throws up a totally random underpinning for the universe. The laws of the universe, long held sacrosanct, turn out to be laws in a statistical or averaged sense. More in spirit than in letter. Consider the time honored Law of Conservation of matter and energy. It works, and we all know it. But it works within the capability of our equipment and observation. Nothing more. This is also true for other such laws. But suppose spacetime is fuzzy and we can observe only so much? In any case, the last word has not been spoken.

At the end of the second millennium, unthinkable questions were raised on such supposedly once and for all closed issues as the Big Bang model, the standard model of particle physics, the constancy of sacrosanct quantities like the fine structure constant, the speed of light, the Gravitational constant and so forth, as dramatic new observations stampeded in.

7.9   Blow out Infinity

The question is, are there any holy cows at all (except possibly in India)? And then many astronomers are asking today: are we being as naive as Alexander, who, as tutored by Aristotle, stood on the northern fringes of India and thought that he had the whole of India, and even the ends of the earth at hand? May be – and all this is speculative – there are many "parallel" or additional universes? In the spirit of the idea of the nineteenth century Swiss physicist Charlier, it appears today, that the building blocks of the universe are not galaxies, but rather clusters of galaxies contained in superclusters of galaxies and so on. May be our universe itself is nothing more than one such building block?

There is one hint though. The universe as we know it appears to be a black hole in a sense, and we are living right inside it! Remember the minimum space time intervals which define the "interior" of particles? We do not know what happens inside these fuzzy intervals – our laws no longer apply. For an observer inside the universe, however, it could well be the reverse! Suppose our universe is a larger counterpart of such a tiny interval-particle? Thus there could be any number of such "universe" black holes or "particles". Only an observer whose measurements could span several life times of the universe would be able to know of their existence! Could our mind-boggling universe be, but a silly "particle"?

To put it another way, during the life time of the universe of about fourteen billion years, it has swollen to a size of some million, billion, billion kilometers. This distance defines a horizon beyond which even our most powerful telescopes are powerless to probe. But is this horizon the boundary where the story ends? Probably not, is the view to which many astronomers are veering. Beyond the horizon of our direct perception, the universe extends to infinity, to any number of trillions of such universes strewn all across. And who knows, there would be any number of planets identical, or almost identical to our own earth, complete with human beings like us, our cosmic clones, and even a carbon copy of yourself. In fact we can do better. Simple calculations reveal that there would be a galaxy identical to ours, one followed by the number of digits in the above radius of our universe expressed in kilometers–that many kilometers away. All the different universes are parallel universes. Multiverse is the name going around these days.

But even this is not the end of the story. According to other ideas, such a conglomeration of parallel universes is what may be called the Level One Multiverse, something that can be thought of as a bubble. According to modern theories, as we saw the big bang triggered off a super fast expansion called inflation. Such an inflation or super fast blow out could well have spawned any number of such bubbles, each bubble being a huge collection of parallel universes. One could think of these bubbles or the Level One Multi- verses as the many individual bubbles gushing out of a bottle of an aerated drink that has just been opened (Cf.Fig.7.3).

What could distinguish the various Level One Multiverses from each other? In fact they could all be very different with very different values of the electric charge, the gravitational constant and what not.

The story does not stop even here! There could be Multiverses here and now! That would be the case if the interpretation of the century old Quantum Theory due to Hugh Everett III is correct. We encountered this earlier on. In fact many physicists think it is so. According to our usual physics, a particle travelling from a point A to a point B is a fairly straightforward event. Quantum Theory however, tells us that the particle could go from A to B by any number of different routes other than what we see. The usual interpretation if you remember is that by the very act of observing the particle move from A to B we eliminate all other possibilities. This has been the generally accepted point of view. But in 1957 Everett as we saw proposed that all other possibilities which have supposedly been snuffed out actually do take place, and its only one of them which we get to see. So in the simple act of, let us say an electron going from one point to another

Fig. 7.3 The universe of universes is but a bubble. Each dot is a universe in itself

point, a centimeter away, there are millions of hidden acts that have taken place, each in its universe. The tiniest bit of space conceals an infinity of mysterious universes!

And this is not all. Einstein's general theory of relativity leads us to the conclusion that when a star collapses into a black hole, at the very center there is what is called a singularity. A singularity can be thought of as a junction or crossroads of infinitely many different roads. Only in our case it is the junction of infinitely many universes each legitimate in its own right, and moreover each with its own laws of nature. This is because at the singularity itself, all laws of nature breakdown, just as, exactly at the

North Pole there is no meaning to the East, West or other directions. Every direction can be the East and so forth [87].

Some scientists have also come to a similar conclusion from the viewpoint of superstring theory you may remember. It is reckoned that there would be so many universes well, one followed by some five hundred zeros! There would be a whole landscape of universes. Why would this be so? Because there are so many possible string theory solutions. And they do not give the correct value of the cosmological constant. The different universes would have different values for these constants, the cosmological constant included. Surely one of those would have the value of our cosmological constant, and that is the universe we inhabit.

In my own model it turns out that the universe can be thought to be a blown up version of an elementary particle which is spinning. This need not end with our universe. Why should it? A huge number of such "particle universes" could form the analogue of a super particle universe—the analogue of let us say, gas molecules in a cubic centimeter. And so on, possibly with increasing dimensionality of space and time at each step. It is a bit like colonies of colonies of colonies and so on. Our own universe might continue expanding for another ten thousand times its present age-and to ten thousand times its present radius.

All this is not a fantasy dreamt out by scientists though it comes pretty close to that description. There are very delicate tests proposed which can provide a clue. There is another interesting aspect. We know that Max Planck introduced his constant at the time of proposing the Quantum theory, so as to explain experiments. There is no fundamental derivation of this constant todate, except for the original motivation. You may ask why should we stop with one constant. Why not other constants? Indeed it is possible to introduce something like the Planck constant, though at different scales of the universe [88]. So there could be a Planck constant for the solar system, one for galaxies, one for super clusters and so on. It is then possible to explain some observations. For instance with a constant for the solar system, we can deduce the distances of the various planets from the Sun. Similarly with the constant for galaxies, it is possible to explain a rather mysterious observation, of more recent times. This is that there are huge voids in the universe. It is not that galaxies are spread out uniformly like molecules in a container. Rather there seem to be shells at different distances and galaxies populate these shells. There are gaps in between the different shells. A some what similar situation can be seen for the energy levels of electrons and atoms or the orbits of planets in the solar system.

Of course as yet there are no fundamental calculations from which these, what you may call scaled Planck constants arise. It's just that the values explain such features of the solar system and the large scale universe. In the words of Sir Martin Rees, Astronomer Royal of England, "Our universe may be even one element-one atom, as it were-in an infinite ensemble; a cosmic Archipelago. Each universe starts with its own big bang, acquires a distinctive imprint (and its individual physical laws) as it cools, and traces out its own cosmic cycle. The big bang that triggered our universe is, in this grander perspective, an infinitesimal part of an elaborate structure that extends far beyond the range of any telescopes." Clearly our fantasy doesn't seem to be able to catch up with the mind blowing developments of science. Interestingly, ancient Indian mythology describes such a scenario. The age of the universe is but a full day in the life of Brahma. Incredibly, the age of the universe is computed to be some 8 billion years – not too far off the mark! But then, Brahma lives, not just for a day, but for hundreds of years! These are some of the challenges which we have to squarely face in the third millennium and find the answers. Or, will the universe carry its dark secrets to its dark grave?

Imagine a person born in an incubator and bred there, without any contact-not even eye contact-with the outside world. Suddenly a window opens out and the person gets the first glimpse of the world at large, but only for a brief moment. The window shuts again. The person is then asked to write out the history and geography of the world. Aren't we humans in a somewhat similar predicament? Hopelessly isolated in time and space, we yet dare to ask questions about the birth, nature and death of the universe itself! Yet as the Late Holy Seer of Sringeri in South India remarked, after an extended Planetarium presentation followed by an encore, the miracle is not so much that we are so small and the universe so large, but rather that we are so small and yet can know so much about the so large.

Bibliography

[1] Sidharth, B.G. (1999). *The Celestial Key to the Vedas* (Inner Traditions, New York).

[2] Koestler, A. (1959). *Sleep Walkers* (Hutchinson, London), pp.247ff.

[3] Srinivasa Iyengar, C.N. (1967). *The History of Ancient Indian Mathematics* (World Press, Calcutta).

[4] Pannekoek, A. (1961) *A History of Astronomy* (Dover Publiations, Inc., New York).

[5] Isaac Asimov (1996). *The Kingdom of the Sun* (Collier Books, New York).

[6] Cooper, L.N. (1968). *An Introduction to the Meaning and Structure of Physics* (Harper International, New York).

[7] Pannekoek, loc.cit.

[8] Bergmann, P.G. (1969). *Introduction to the Theory of Relativity* (Prentice-Hall, New Delhi), p.248ff.

[9] Einstein, A. (1965). *The Meaning of Relativity* (Oxford and IBH, New Delhi), pp.93–94.

[10] Pannekoek, loc.cit.

[11] George Gamow (1962). *Gravity* (Doubleday and Company Inc., New York).

[12] Wheeler, J.A. and Tegmark, M. (2001). *Hundred Years of Quantum Mysteries* in *Scientific American* February 2001.

[13] Kuhn, T. (1970). *The Structure of Scientific Revolution*, 2nd. ed. (Chicago University Press, Chicago).

[14] Rae, A.I.M. (1986). *Quantum Mechanics* (IOP Publishing, Bristol), pp.222ff.

[15] Wheeler, J.A. (1968). *Superspace and the Nature of Quantum Geometrodynamics, Battelles Rencontres, Lectures*, De Witt, B.S. and Wheeler, J.A. (eds.) (Benjamin, New York).

[16] Pannekoek, loc.cit.

[17] Cooper, L.N., loc.cit.

[18] Elias, V., Pati, J.C. and Salam, A. (1977). *Centre for Theoretical Physics, University of Maryland, Physics Publication*, September 1977, pp.78–101.

[19] Gleick, J. (1987). *Chaos: Making a New Science* (Viking, New York).

[20] Wheeler, J.A. and Feynman, R.P. (1945) *Rev.Mod.Phys* 17, p.157.

[21] Sidharth, B.G. (2001). *Chaotic Universe: From the Planck to the Hubble Scale* (Nova Science, New York).

[22] Singh, V. (1988). *Schrodinger Centenary Surveys in Physics*, Singh, V. and Lal, S. (eds.) (World Scientific, Singapore).

[23] David Z Albert and Rivka Galchen (2009). *Was Einstein Wrong?: A Quantum Threat to Special Relativity* in *Scientific American* February 2009.

[24] In ancient Indian thinking there are vast spans of time each equalling approximately the age of the universe. However our universe lives but for one day in the life of Brahma. Brahma himself has a very long life span. The concept is therefore not just of chance but also of what today we may call parallel universes.

[25] Hugh Everett III was a student of J.A. Wheeler and proposed what came to be known as the many universes theory. At each point of space there is a different universe. We live and evolve in one of them. In this connection Cf. Max Tegmark, Parallel Universes, Scientific American, April 2003.

[26] Hawking S.W. (1992). *A Brief History of Time* (Bantam Press, London).

[27] Toffler, A. (1985). in *Order Out of Chaos* (Flamingo, Harper Collins, London).

[28] George G. Szpiro (2007). *Poincare's Prize: The Hundred-Year Quest to solve one of Math's Greatest Puzzle* (Penguin, USA).

[29] Nicolis, G. and Prigogine, I. (1989). *Exploring Complexity* (W.H. Freeman, New York), p.10.

[30] Wheeler, J.A. (1994). *Time Today* in *Physical Origins of Time Asymmetry* (Eds.J.J. Halliwell, J. Perez-Mercader, W.H.Zurek) (Cambridge University Press, Cambridge).

[31] Hawking, S.W. and Werner Israel (1987). *300 Years of Gravitation* (Cambridge University Press, Cambridge).

[32] Wheeler, loc.cit.

[33] Misner, C.W., Thorne, K.S. and Wheeler, J.A. (1973). *Gravitation* (W.H. Freeman, San Francisco), pp.819ff.

[34] Prigogine, loc.cit.

[35] Sidharth, B.G. (2008). *The Thermodynamic Universe* (World Scientific, Singapore).

[36] Prigogine, I. (1997). *The End of Certainty* (Free Press, New York).

[37] Nandalas Sinha (trans.) (1986). *Vaiseshika Sutras of Kanada* (S.N. Publications, Delhi).

[38] Frisch, D.H. and Thorndike, A.M. (1963). *Elementary Particles* (Van Nostrand, Princeton), p.96ff.

[39] Frisch and Thorndike, loc.cit.

[40] Over the past few centuries a methodology of science has evolved. This has been time tested. Thus initial observations led to the framing of a hypothesis. This hypothesis then must be tested for its veracity. Those hypothesis which are either provable or disprovable are considered to be meaningful. Further there must be a minimum of hypotheses which explain a maximum of observations, including those in the future. This also implies that the hypothesis that is simplest is prefered over more complicated hypotheses. A textbook example of this method is Einstein's special theory of relativity. The observed Michelson-Morley experimental results were individually

explained by Poincare on the one hand and Lorentz and Fitzgerald on the other. Einstein's single explanation could explain both these diverse results. Moreover the predictions of the special theory of relativity were subsequently confirmed by experiments.

[41] Salam, A. (1990). *Unification of Fundamental Forces* (Cambridge University Press, Cambridge). This book gives a delightful account of steps leading to electroweak unification.

[42] Salam, loc.cit.

[43] G 't Hooft. (2008). in *A Century of Ideas* Sidharth, B.G. (ed.) (Springer, Dordrecht).

[44] Gullick, D. (1997). *Encounters With Chaos* (McGraw Hill, New York), p.114ff.

[45] Gleick, loc.cit.

[46] The extremely sensitive dependence on what are called initial, or loosely external conditions orf chaotic systems has a well known metaphor: The flapping of the wings of butterflies in South America can cause changes in the rain patterns in Africa. This has come out of the work of Conrad Lorenz.

[47] Nicolis and Prigogine, loc.cit.

[48] Mandelbrot, B.B. (1982). *The Fractal Geometry of Nature* (W.H. Freeman, New York), pp.2,18,27.

[49] Dyson, F.J. (1988). *Infinite in all directions* (Harper and Row, New York).

[50] Mandelbrot, loc.cit.

[51] Prigogine, loc.cit.

[52] Lucas, J.R. (1984). *Space Time, And Causality* (Oxford Clarendon Press).

[53] Nambu, Y. (1981). *Quarks - Frontiers in Elementary Particle Physics* (World Scientific, Singapore) pp.212.

[54] Around 1930 Dirac turned to work out the model of an atom which is more like a shell or membrane. Though much work was done, this model led to increasing mathematical and physical difficulties and was finally abandoned.

[55] Greene, B. (1999). *The Elegant Universe* (Vintage, London).

[56] Sidharth, loc.cit.

[57] Witten, E. (1996). *Physics Today* April 1996, pp.24–30.

[58] Peat, F.D. (1988). *Super Strings* (Abacus, Chicago), p.21.

[59] Smolin, L. (2006). *The Trouble with Physics*, (Houghton Mifflin Company, New York).

[60] Smolin, loc.cit.

[61] Sidharth, B.G. (2007). *Encounters:Abdus Salam* in *New Advances in Physics* Vol.1, No.1, March 2007, pp.1–17.

[62] Laughlin, R.B. (2005). *A Different Universe* (Basic Books, New York).

[63] Hawking, loc.cit.

[64] Weinberg, S. (1979). *Phys.Rev.Lett.* 43, pp.1566. Also Weinberg, S. (1993). *Dreams of a Final Theory* (Vintage, London).

[65] It appears that Zelmanov in the erstwhile Soviet Union had worked out in detail the anthropic principle, or a version of it and Hawking who had visited Zelmanov seemed to have been impressed with the idea. (Cf. also D. Rabounski, Progress in Physics, Vol.1, January 2006).

[66] Peter Woit (2006). *Not Even Wrong* (Basic Books, Cambridge).

[67] In certain schools of Indian thought, particularly those related to Buddhism, there was this philosophy of momentarius or "kshanabhangabada". According to this philosophy the universe is momentary. It gets destroyed every instant to be recreated the next instant.

[68] Wheeler, J.A. (1983). *Am.J.Phys.* **51** (5), May 1983, pp.398-404.

[69] Sidharth, loc.cit.

[70] Dirac, P.A.M. (1958). *The Principles of Quantum Mechanics* (Clarendon Press, Oxford), pp.4ff, pp.253ff.

[71] Weinberg, S. (1972). *Gravitation and Cosmology* (John Wiley & Sons, New York), p.61ff.

[72] Hooft, G.'t. (1989). *Gauge Theories of the Forces between Elementary Particles* in *Particle Physics in the Cosmos*, Richard A Carrigan. and Peter Trower, W. (eds.) *Readings from Scientific American Magazine* (W.H. Freeman and Company, New York).

[73] Wheeler, loc.cit.

[74] Feynman, R.P. (1985). *QED* (Penguin Books, London).

[75] Mandelbrot, loc.cit.

[76] Sidharth, B.G. *Dark Energy and Electrons Int.J.Th.Phys.* (In press).

[77] Glinka, L.A. (2009). *Apeiron* April 2009.

[78] Sidharth, B.G. *Mass Generation and Non Commutative Spacetime* in *Int.J.Mod.Phys.E.* (In press).

[79] Sidharth, B.G. (2005). *The Universe of Fluctuations* (Springer, Netherlands).

[80] Lloyd Motz (1980). *The Universe its beginning and end* Abacus, London).

[81] Sidharth, B.G. (2008) *A Century of Ideas* (Ed.B.G. Sidharth) (Springer, New York).

[82] Coughlin, G.D. and Dodd, J.E. (1991). *The Ideas of Particle Physics* (University Press, Cambridge).

[83] Davies, P. (1989). *The New Physics* Davies, P. (ed.) (Cambridge University Press, Cambridge).

[84] The accelerated expansion of the universe was deduced on the basis of a careful observation of Type II Supernovae. A number of researchers were involved including S. Perlmutter and R. Kirshner. These results were revealed at the American Association of Astronomy Meet in 1998.

[85] Wilczek, F. (1999). *Physics Today* January 1999.

[86] Bronowski, J. (1973). *The Ascent of Man* (BBC, London).

[87] Michio Kaku and Jennifer Trainer (1987). *Beyond Einstein The Cosmic Quest for the Theory of the Universe* (Bantam Books, London).

[88] Sidharth, loc.cit.

Index

't Hooft, 120, 138, 149
(LHC), 1

A.H. Compton, 39
Abdus Salam, 19, 43, 88, 91, 124, 157
Abhay Ashtekar, 107
aether, 161, 162
Africa, 27
Alain Aspect, 132
Alexander, 171
Alexandria, 5
Alfonso Rueda, 147
alpha particles, 34
Amelino-Camelia, 149
Ampere, 14, 87
Anaximenes, 4
Andromeda galaxy, 22, 57
anti matter, 81
Antonio Damasio, 74
Antonio Machado, 127
anu, 76
Arabs, 6
Aristarchus, 6
Aristotle, 8, 171
Arno Penzias, 154
Arp, 154
Atomic Theory, 121
Australia, 164

Baghdad, 6
Balmer series, 32
Belgium, 140

Bell Telephone Laboratories, 154
Bergson, 130
Berkeley, 82
Bernhard Hirsch, 147
Bhagawad Gita, 151
Bhagvad Gita, 159
Big Bang, 93, 153, 154, 157, 158, 164, 171
black hole, 145
boson, 90, 105
Bosonic String, 104
Brahma, 175
Brahmagupta, 6
Brans, 164
Brownian motion, 98
Burton Richter, 118

C-field, 153
C. Lattes, 83
C. Wiegand, 82
C.D. Anderson, 81
C.R.T. Wilson, 85
Cambridge University, 31
Carlo Rovelli, 107
Casimir effect, 162
CERN, 1, 102, 123, 169
Chamberlain, 82
Charlier, 171
Cherenkov radiation, 85
Christian Huygens, 161
Clifford, 99
Columbia University, 115

Compton scale, 112, 135, 138, 139, 146
Compton time, 134
Compton wavelength, 134, 139, 141, 146
Copenhagen, 36
Copenhagen interpretation, 49
Copernicus, 5
Cosmic Microwave Background, 159
cosmological constant, 100, 163
cosmology, 167
Coulomb, 13
Cowan, 88

D. Glaser, 85
Dark Energy, 145, 158, 160
dark energy, 146
dark matter, 93, 157
David Gross, 114
Davisson, 37
De Broglie, 37, 39, 87
Democritus, 77
Dick, 165
dinosaurs, 121
Dirac, 79, 87, 102, 106, 113, 134, 165, 167
Dirac equation, 41, 79, 149
Double Slit experiment, 45
Dyson, 97

E Sagre, 82
Eddington, 27, 59, 140, 150, 166
Edinburgh, 85
Edward Nelson, 138
Edward Teller, 81
Edward Witten, 116
Edwin Hubble, 167
Edwin P. Hubble, 151
Egyptians, 4
Einstein, 4, 9, 11, 18, 19, 22, 25, 33, 66, 87, 99, 142, 151, 170
Einstein-Rosen-Podolsky paradox, 50
electrodynamics, 18
electromagnetic, 88
electromagnetic radiation, 36

electromagnetism, 25, 87, 88, 91, 93, 104
electron, 87–90, 102, 145
Electrons, 45
elementary particles, 91, 93, 102
Empire State Building, 29
Empire State building, 29
England, 77, 81
Entanglement, 54
Entropy, 58
EPR paradox, 53, 54, 132
Ernest Rutherford, 35
Ernst Mach, 168
Euclid, 97
Eugene Wigner, 71, 170

F.J. Dyson, 87
Faraday, 87
fermion, 90, 105
Fibonacci, 6
Financial Times, 119
Fitzgerald, 17
Foucalt, 12
France, 77, 133
Fred Hoyle, 152
fuzzy spacetime, 170

G. Veneziano, 102
Galilean Relativity, 13
Galileo, 7, 11, 23
Gamow, 29
Geiger counter, 34, 45, 85
General Relativity, 71, 142, 144
general relativity, 93, 106, 107, 109, 170
General Theory of Relativity, 87, 152
Geneva, 75, 102
Georg Cantor, 97
George Gamow, 153, 155
Georgi, 91
Gerard 't Hooft, 92
Germany, 77
Germer, 37
Glashow, 91
GLAST, 1, 148
gluons, 91

Grand Unified Force, 91
Gravitation, 24
gravitation, 86–88, 93, 101, 104–107, 109, 168
graviton, 90
Graz, 127
Greek, 6, 77
Greek atomism, 121
Greeks, 4
Greisen, 148
Grossman, 18

H Dieter Zeh, 55
H.E. Puthoff, 147
Hans Bethe, 153
Hartland Snyder, 137
Hegel, 130
Heidegger, 130
Heisenberg, 99
Heisenberg Uncertainty Principle, 70, 161
Heisenberg Uncertainty principle, 135
Helium, 34
Heraclitus, 130
Hermann Bondi, 152
Hermann Weyl, 94, 140
Hideki Yukawa, 83
Higgs, 149
Higgs boson, 91
Higgs bosons, 169
Holland, 17
Hubble, 152
Hugh Everett III, 55, 172
Huygens, 34

Ilya Prigogine, 72
India, 76, 171
inertia, 8

J.J. Thompson, 78
James Chadwick, 78
Jean Perrin, 98
Jogesh Pati, 91
John Dalton, 77
John Wheeler, 93
Josephson, 112

Jupiter, 15, 127

Kaluza, 95, 104, 107
Kanada, 4, 75, 77
Kanka, 6
Kepler, 6, 11, 87, 127
Kerr-Newman black hole, 145
Khalif Al Mansoor, 6
King Oscar II, 60
Kirshner, 159
Klein, 104
Kuzmin, 148

Laplace, 96
Large Hadron Collider, 169
Large Hadron Collidor, 75, 124
Laughlin, 112, 113
Lavosier, 77
Laws of Motion, 87
Lee Smolin, 107, 109, 115, 116
Leibnitz, 144
leptons, 89, 91
LHC, 75, 124, 169
Liebniz, 100
Loop Quantum Gravity, 99, 107, 109, 144, 170
Lord Kelvin, 121
Lord Rayleigh, 33
Lorentz, 17–19, 102
Los Alamos, 55
Lucippus, 77
Ludwig Boltzmann, 63

M Theory, 118
M-Theory, 107
M. Deutsch, 81
magnetism, 87, 91
Maitri Upanishad, 57
Mandelbrot, 97, 98
Manhattan, 22
Marcell Grossmann meet, 159
Mars, 6, 127
Martinus Veltman, 149
mass, 90–92, 94, 101, 105, 106, 167
Max Born, 40
Max Planck, 33, 78, 95, 174

Maxwell, 14, 87
mechanics, 25
Meissner, 112
membrane, 102, 106
membranes, 106
Mendelev, 77
Mercury, 128, 165
Michelson, 15, 17, 161
monopole, 92, 93
Morley, 15, 17, 161
MRI, 43
Multiverse, 172

Narlikar, 154
NASA, 1, 148, 166
Neils Bohr, 35, 49
Netherlands, 140
neutrino, 44, 88, 89, 91, 92, 94
Neutrino Spin, 89
neutrons, 21, 88
New Yorker magazine, 81
Newton, 11, 13, 14, 22, 58, 87, 96,
    129, 144, 151
Newtonian, 100
North Pole, 12, 70, 174

Occam's razor, 86
Oersted, 13
Old Quantum Theory, 36
Oskar Klein, 94
Otto Stern, 42

P.A.M. Dirac, 41
P.M.S. Blackett, 81
Parmenides, 130
Parvati, 53
Paul Dirac, 159
Pauli, 88, 93
Pauli Exclusion Principle, 79
perfect solids, 127
perihelion, 165
Perimeter Institute, 115
Perlmutter, 159
Perrin, 129
Peter Higgs, 75
Peter Woit, 115, 120

phase transition, 91
photoelectric effect, 33
photons, 90, 91
Physics Today, 120
Physics World, 160
pi-meson, 102
pion, 91
Pioneer 10, 166
Pioneer 11, 166
Planck, 107
Planck energy, 125
Planck length, 109, 142, 154
Planck scale, 138, 146
Planck springs, 146
Planck time, 146
Plato, 4
Podolsky, 52
Poincaré, 17–19, 43, 59
Poincare, 102
Portugal, 140
Prigogine, 97, 99, 129, 138
proton, 88–91
protons, 44
Prout, 78
Ptolemy, 5
pulses, 166
Pythagoras, 149

QED, 90
QFT, 44
quantum, 36
Quantum computers, 54
Quantum Field Theory, 43, 100, 130,
    137, 144
Quantum fluctuation, 160
Quantum Gravity, 96, 155
Quantum mechanical experiment, 47
Quantum mechanics, 42, 43, 79, 87,
    98, 161
Quantum super string, 170
Quantum superstring theory, 106
Quantum superstrings, 96, 104
Quantum Theory, 87
Quantum theory, 39, 48, 56, 58, 79,
    93, 98, 101, 104, 107, 130, 131, 160,
    170

Quantum Vacuum, 158
Quantum vacuum, 160–162
quark, 89, 91
Quasars, 164

R.B. Laughlin, 111
Ralph Alpher, 153
Random Walk, 72, 167
Regge, 102
Regge trajectories, 104
Relativity, 18
Renormalization, 144
Richard Feynman, 46, 63, 111, 144
Richardson, 97
Richter, 77
Rig Veda, 3, 163
Ritz Combination Principle, 36
Ritz combination principle, 32
Robert Wilson, 154
Roger Penrose, 64
Roman numerals, 6
Romer, 15
Rosen, 52
Russia, 77
Rutherford, 78

satellite, 68
Saturn, 127
Schrodinger, 39, 48
Schrodinger's cat, 48, 56
Scientific American, 117, 160
Sean Carroll, 116
Second Law of Thermodynamics, 58
Seer of Sringeri, 175
semi conductors, 43
Sheldon Glashow, 88, 120
Sir Isaac Newton, 7
Sir James Jeans, 33
Sir Karl Popper, 121
Sir Martin Rees, 175
Sir W.H. Bragg, 39
Siva, 53
SLAC, 123
Sloane Digital Sky Survey, 160
Solvay Conference, 114, 115

spacetime, 71, 87, 93, 99, 100, 104,
    106, 107, 109, 144
Spain, 6, 140
special relativity, 87, 102, 109
Special Theory of Relativity, 20
spin, 90, 91, 93, 94, 105, 106
Standard Linear Accelerator Center,
    118
standard model, 91–93
Steady-State model, 153
Stephen Hawking, 70, 117
Steven Weinberg, 88, 159
String Theory, 112
strings, 106, 107
strong force, 88, 89, 91, 92
super conductors, 43
Super Kamiokande, 92
supersymmetry, 105, 106
SUSY, 105–107
Sweden, 60
Switzerland, 18
symmetry, 91

T. Regge, 101
T. Ypsilantis, 82
T.D. Lee, 137
tachyons, 104
Ted Jacobson, 107
Theodore von Kaluza, 94
thermodynamics, 32, 59
Thomas Gold, 152
Thomas Kuhn, 31
Thompson, 38
Time Magazine, 115
Toffler, 58
Tommy Gold, 153
total solar eclipse, 27
Tycho Brahe, 6, 87

Ujjain, 6
Ultra High Energy Cosmic Rays, 148
Uncertainty Principle, 40
unification, 87, 88, 91, 106, 109
United States, 81
universe, 88, 92, 100, 105, 107
University of California, 82

University of Chicago, 116

V.L. Ginzburg, 141
V.M. Slipher, 152
Vedas, 4
Veneziano, 103
Venus, 128
Vishnu Purana, 77
Von Klitzing, 111

W.H. Zureck, 55
Walter Gerlach, 42
wave equation, 39
weak interaction, 44
Webb, 164
Weinberg, 136
Werner Heisenberg, 40

Wheeler, 66, 71, 93, 144
Wilzeck, 164
WIMPS, 157
Witten, 105
WMAP, 160
Wolfgang Pauli, 44, 80, 116

X-rays, 39

Y. Nambu, 83
Ylem, 155
Yukawa, 83

Zatsepen, 148
Zero Point Field, 158, 160, 162
Zitterbewegung, 135, 143, 146
Zwicky, 157